U0143286

[英国] 杰奎琳·斯特多尔 著　张弢 译

数学简史

牛津通识读本·

The History of Mathematics

A Very Short Introduction

译林出版社

图书在版编目（CIP）数据

数学简史 ／（英）杰奎琳·斯特多尔（Jacqueline Stedall）著；张弢译．
—南京：译林出版社，2023.10
（牛津通识读本）
书名原文：The History of Mathematics: A Very Short Introduction
ISBN 978-7-5447-9865-5

Ⅰ.①数… Ⅱ.①杰… ②张… Ⅲ.①数学史－世界－普及读物 Ⅳ.①O11-49

中国国家版本馆 CIP 数据核字（2023）第 165186 号

The History of Mathematics: A Very Short Introduction by Jacqueline Stedall
Copyright © Jacqueline Stedall 2012
The History of Mathematics: A Very Short Introduction, First Edition was originally published in English in 2012.
This licensed edition is published by arrangement with Oxford University Press.
Yilin Press, Ltd is solely responsible for this bilingual edition from the original work and Oxford University Press shall have no liability for any errors, omissions or inaccuracies or ambiguities in such bilingual edition or for any losses caused by reliance thereon.
Chinese and English edition copyright © 2023 by Yilin Press, Ltd
All rights reserved.

著作权合同登记号　图字：10-2018-429 号

数学简史　［英国］杰奎琳·斯特多尔 ／ 著　张　弢 ／ 译

责任编辑　陈　锐
特约编辑　茅心雨
装帧设计　景秋萍
校　　对　孙玉兰
责任印制　董　虎

原文出版　Oxford University Press, 2012
出版发行　译林出版社
地　　址　南京市湖南路 1 号 A 楼
邮　　箱　yilin@yilin.com
网　　址　www.yilin.com
市场热线　025-86633278
排　　版　南京展望文化发展有限公司
印　　刷　江苏凤凰通达印刷有限公司
开　　本　890 毫米 ×1260 毫米　1/32
印　　张　8.375
插　　页　4
版　　次　2023 年 10 月第 1 版
印　　次　2023 年 10 月第 1 次印刷
书　　号　ISBN 978-7-5447-9865-5
定　　价　39.00 元

版权所有·侵权必究
译林版图书若有印装错误可向出版社调换．质量热线：025-83658316

序　言

田　淼

非常荣幸有机会向读者介绍杰奎琳·斯特多尔的这本《数学简史》。以一百余页的篇幅完成对数学史的总体介绍，几乎是一个不可能完成的工作，然而，此书不仅做到了，还令人耳目一新，且展现了数学史研究的新视角，给人以新的启迪。

此书的结构与通常所见的按时序或按领域分支勾勒数学发展的通史性著作完全不同，而是选取了“什么是数学与数学家”“数学思想的传播”“数学的学习”“数学家如何谋生”等问题进行综合性的介绍与论述，并就其中一些至关重要的案例做了阐释。

本书的第一章以广受注目的费马大定理的研究历史为切入点，作者通过对相关数学家的分析，对三类数学史研究展现的图景提出质疑。“象牙塔”版本的数学论述方式忽略了数学家所处的社会与环境；只关注相关重要成果的“垫脚石”类的研究方式展现了不同阶段的研究高峰，但重要成果之间的工作和努力则无法体现；“精英”版本的历史图景则无视了这些伟大人物周边的人的工作。在此后的章节中，斯特多尔给出了与这些传统研究模式不同的视角。她将关注点更多地放在数学知识和数学家在不同地域和不同时间段的具体构成及其存在的方式，数学知识和思想如何实现传播以及数学史研究中如何理解、翻译及阐释历史文献等问题上。在此研究中，她对中国及其他非欧洲地区的数学内容和研究方式做了较为细致的探讨，以展现数学本身的多样性、地域性和时代性特征。

作为《英国数学史公报》的编辑及《牛津数学史指南》的主编之一，斯特多尔的研究视野宽广，并与众多数学史工作者有着密切的联系，这无疑是她能够完成这一著作的基础。《数学简史》自出版以来便得到了数学史界的广泛好评，杰奎琳·斯特多尔因此书于2013年获得英国数学史学会的诺伊曼奖（以英国数学家彼得·诺伊曼命名），其获奖评语评价该书"具有启发性，写作上乘，且非常适合大众读者，同时也包含很多新的及有洞见的评论"。

对数学专业学生来说，此书有助于理解数学的内容及其所从事的行业在不同地区和不同发展阶段的特征，以加深对自己领域的历史性认识。由于书中仅有很少的数学公式，它对于非数学专业的读者也具有很强的亲和性，此书并非数学发展的线性描述，而是就与数学相关的一些重要问题进行深入探讨，书中包含丰富的历史知识，并对以往数学史研究所忽视的公众对数学的理解与数学发展的关系等有着妙趣横生的描述，读来引人入胜。

值得强调的是，对于数学史及科技史研究者，此书提供了新的视角，其中对于数学多样性的描述及如何在社会、文化与环境中认识和理解数学的发展和数学家的身份特征等问题的论述，具有方法论上的启发性。此外，《数学简史》中，作者并没有为她提出的问题给出标准性答案，可以说，书中的很多问题都是开放性的，读者可以进行自己的思考并给出各自的答案。与数学一样，数学史研究也具有很强的多样性，正是由于其多样性，数学史及科学史才能够长盛不衰且异彩纷呈。为此，我恳请我的同行们在阅读过程中，暂时忽略诸如《九章算术》的成书年代和过程等学术界仍有争议的问题，而将注意力更多地集中于其中蕴含的深入思考与研究方法。

目 录

致谢

我在给这样一个宏大主题撰写通识读本的过程中，受到了该系列其他作者极大的启发。他们中的许多人尽力以富于想象而又引人思考的方式写作，接受了同样严苛的挑战。

在过去几年中，我有幸编辑了《牛津数学史手册》以及英国数学史学会的期刊《英国数学史公报》。这使我与八十多位作者建立了密切的工作关系，他们从各种视角来写数学史。我从他们每人身上都学到了些东西。其中大部分工作都是我与埃莉诺·罗布森一起完成的，在我的朋友和同事中她是最好的一位。我非常感谢与她一起共事和讨论的时光，那段时间我试图在本书中传达的一切渐渐成形。我特别借鉴了以下诸位的研究和专业知识：马库斯·阿斯珀、索尼娅·布伦特延斯、克里斯托弗·卡伦、马里特·哈特维特、安妮特·伊姆豪森、基姆·普洛夫克尔、埃莉诺·罗布森、科琳娜·罗西、西蒙·辛格、波莉·特内勒基和本杰明·沃德豪；这些作者及其他一些人的书籍和文章，可以在书末建议的延伸阅读书单中找到。

约翰·赫尔西所收藏的孩童抄本（在第四章中论及）是数

学协会的财产，收藏在位于莱斯特大学的戴维·威尔逊图书馆中。我对协会的档案管理员玛丽·沃姆斯利和麦克·普赖斯在我研究过程中给予的热情款待与合作表示感谢。我也感谢牛津大学伍斯特学院的乔安娜·帕克，她让我看到了约翰·奥布里抄录的安妮·埃特里克的笔记本。我要感谢安德鲁·怀尔斯、克里斯托弗·卡伦、埃莉诺·罗布森及亚当·西尔弗斯坦，他们不辞劳苦地检查了第一、二、四、五章中的细节。我向他们以及就本书的各个方面发表过机敏评论的所有其他人表示由衷的感谢：牛津大学出版社不知名的诸位读者，连同彼得·诺伊曼、哈维·莱德曼、杰西·沃尔夫森以及我直系亲属中的所有成员，他们中的一些人直到现在都从未想过要了解数学的历史。

导　言

数学的历史可以追溯到至少四千年前，而且渗透进每一种文明与文化之中。即便是在像本书这样非常精简的介绍中，也可以大致按时间顺序来概述一些关键的数学事件和发现。确实，这或许正是大多数读者所期望的。不过，那样的阐述可能会有几个问题。

首先，这种叙述往往会刻画出一种辉格党版本的数学史。在这样的版本中，人们通常认为数学理解是朝着如今的辉煌成就径直地发展而来的。不幸的是，那些找寻前行证据的人士往往忽视了其中的复杂性、失误及绝望，而这些正是任何一种人类的努力都无法避免的，数学也是其中之一；有时失败可以和成功一样有启迪作用。此外，通过把当下的数学定义为用来衡量早前努力的基准，我们就会很轻易地将过去的贡献视作勇敢坚决却最终过时了的努力。相反，在试着去了解这个或那个事实或定理是如何起源的时候，我们需要将那些发现置于其自身所处的时空背景下来审视。

第二个问题是，按时间顺序排列的记述通常都遵循“垫脚石”的风格。在这样的叙述中，一个个发现被摆在我们面前，而它们之间并没有什么至关重要的联系。我会在后面进一步谈到这个问题。历史学家的目标，不仅是编制事件的时序列表，而且要阐释那些导致事件发生的影响及相互作用。这将是本书中反复出现的主题。

第三个问题是，关键事件与发现会同关键人物相关联。进一步来说，在大部分数学史上，这些人中的大多数会从16世纪左右开始出现并生活在西欧，而且多是男性。这倒不一定反映了数学史家们以欧洲为中心或者持性别歧视的态度。自文艺复兴以来的欧洲男性文化中，数学的迅速发展致使大量的材料出现，而历史学家也有正当理由认为它们值得研究。此外，我们手上有大量这一时期来自欧洲的资料，而中世纪前欧洲、中国、印度或美洲的资料相对来说就少得多。幸运的是，出自上述区域中一些地方的资料，正开始变得更便于获得且更易于查阅。不过，事实仍在于，对重大发现的关注将人类大多数群体的数学经验排除在外，比如女性、孩童、会计人员、教师、工程师、工厂工人等等；而这种情况通常遍及多个大陆且跨越数个世纪。这显然不合适。在某些显著突破的价值（本书将以其中一个作为开端）不被否认的前提下，必须从许多实践数学之人的角度来思考历史，而不是仅从少数人的观点出发。

对于数学史上大多数叙述中男性意识下的偏见，本书能做的纠正不多；不过，对于欧洲以外其他大陆的数学发展，本书不会只停留在泛泛之谈上；并且本书将尝试探索，那些名字永远不会出现在标准历史中的人士如何且在何处以及为什么从事数学实践。但这样一来，就需要不同于常规时序的考查。

我所计划遵循的替代模式将围绕主题而不是时期来构建。每章都将关注两三个案例研究；如此选择，并不是因为它们在任何方面都是全面或详尽的，而是希望它们能引出想法、问题以及新颖的思考方式。同时，为达成上述理念，我已试着尽可能地揭示出各个故事之间的差异或相似之处，以便读者可以对数学悠久历史中的至少若干方面建立起一个相互关联的观点。我的目的不仅在于展示专业历史学家如今都是怎样对待其专业领域的，而且也要展示非专业人士可以如何思考数学的历史。

通过这种方式，我希望本书能够帮助读者认识到贯穿人类历史的数学活动的丰富性与多样性；并且这将是一份非常精简的介绍，不仅是对过往岁月中的一些数学成就，而且是对数学本身作为一门现代学术性学科的历史的介绍。

第一章

数学：传奇与历史

古老而又难以解决的数学问题成为新闻并不常见。不过在1993年，英国、法国及美国的报纸都报道说，一位四十岁的数学家安德鲁·怀尔斯在剑桥艾萨克·牛顿研究所的一次演讲中展示了对三百五十年间悬而未决的费马大定理的证明。事实证明，这一声明为时尚早：怀尔斯两百页的证明中有一个错误，得花些时间来修正，而两年之后这一证明便严密了。怀尔斯用了九年时间来研究该定理。当时，这个故事成了一本书与一部电视电影的主题，怀尔斯在其中谈到自己最后的突破时热泪盈眶。

数学的这段历史之所以如此吸引公众的想象力，一个原因无疑是怀尔斯本人的形象。他在剑桥这次演讲之前，几乎与世隔绝地工作了七年，一心一意地致力于定理背后深层与复杂的数学原理。然后便是这样一个故事：一位孤独的英雄克服万难，为的是实现一个难以捉摸的目标。而那些在西方文化的神话故事中成长起来的人对此情节已经很熟悉。故事的背后甚至还有一位公

主：只有怀尔斯的妻子知道丈夫的终极目标，并且成为第一个收
1 到完整证明的人。怀尔斯把它当作生日礼物送给了妻子。

第二个原因是定理本身很容易陈述，尽管全球范围内也许不超过二十个人能完全理解费马大定理的最终证明。怀尔斯十岁时就已经对它产生了兴趣，甚至那些早把大部分学过的数学内容忘掉的人，也能理解定理在说些什么；我们稍后会来探讨这条定理。

不过在此之前，请注意在本章第一句中已经提及的三个人：怀尔斯、牛顿，还有费马。这在数学领域是很典型的：数学家通常会用他们中的一位来命名定理、猜想或体系。这是因为大多数数学家敏锐地意识到，他们总是以前辈或同事们所做的工作作为前行的基础。换句话说，数学生来便是一门与历史有关联的学科，其中前人的心血很少被忽视。为了开始思考数学史学家所提出的问题，不妨让我们将费马大定理从1993年的剑桥演讲厅回溯到其更遥远的起点。

费马与他的定理

皮埃尔·德·费马生于1601年，在法国南部度过了他的一生。他是一位受过训练的律师，曾担任图卢兹议会的顾问，那是一个所辖范围很大的司法机构。费马利用所剩无几的业余时间从事数学研究，并且远离巴黎的知识分子活动圈，几乎完全是独自一人在做那些工作。17世纪30年代，他通过巴黎最小兄弟会的修士马兰·梅森与更远地方的数学家们进行交流。但在17世纪40年代，随着身上政治压力的增加，他抽身而出，再次独自进行数学研究。费马取得了17世纪早期数学领域内的一些最

深刻的成果，不过对于其中的大部分，他只乐意用吊人胃口的方式提及一点点。他一次又一次地向他的专题通讯员保证，如果 2
有足够的闲暇时间，他会补充细节；可这种闲暇从来也没出现过。他有时会简单地陈述自己的发现，或者会发出挑战，直白地给出他正思考着的那些想法，却不透露他那些好不容易得来的结果。

他对大定理的第一次暗示便出现在这样的一次挑战中。1657年，他把信件寄给了英国数学家约翰·沃利斯和威廉·布龙克尔。他们没看出他在说些什么，并且认为若做出回应便失了身份而对此不予理会。费马过世后，其子塞缪尔在编辑他的一些笔记和论文时，该定理的完整陈述才为人所知。费马将它记在了他手上丢番图所著的《算术》一书的空白处。我们在适时回顾书中哪些内容启发了费马之前，需要简要地介绍一下费马大定理本身。

回想起上学时所学的数学，几乎每个人都会提到的便是毕达哥拉斯定理。该定理指出，直角三角形最长一边（即斜边）的平方，等于两条短边（即两腰）的平方和。大多数人可能还记得，如果两条短边分别为3及4个单位长度，那么长边将为5个单位长度，因为$3^2 + 4^2 = 5^2$。这种三角形被称作“3－4－5三角形”。有了它，人们借助一根绳子就可方便地在地上画出直角；而教科书编著者也会用到这种三角形，他们想设置无须借助计算器便可解答的问题。

由三个整数构成而满足相同关系的集合还有许多，例如很容易验算$5^2 + 12^2 = 13^2$，再如$8^2 + 15^2 = 17^2$。这样的集合有时记作（3, 4, 5）、（5, 12, 13），等等。它们被称为“毕达哥拉斯三元

组”，而且有无数个这样的三元组。就像数学家喜欢做的那样，现在假定我们稍微改变一下条件，看看会发生什么。如果不是
3 取每个数的平方，而是取它们的立方，又会怎么样？我们能否找到使$a^3+b^3=c^3$成立的三元组（a, b, c）呢？或者我们能否更进一步来找寻一个三元组，使得$a^7+b^7=c^7$甚至$a^{101}+b^{101}=c^{101}$成立？费马得出的结论是尝试毫无意义：对于任何超过平方的幂运算，我们找不到使等式成立的三元组。不过，和往常一样，他把处理细节的工作留给了其他人。这一次，他的借口不是时间问题，而是空间不够：他说找到了一个绝妙的证明，只是页边的空白处太小，写不下了。

丢番图的《算术》一书有个1621年的版本，是克洛德·加斯帕尔·巴谢所译。费马就在这样一个译本第八十五页的空白处记下了上述问题。自从1462年在威尼斯重新发现了一份希腊文手稿以来，《算术》一书就一直吸引着欧洲数学家。对于丢番图本人，人们过去一无所知，而现在所了解的就更少了。这份手稿称他为“亚历山大的丢番图”，因此我们可以假设他在埃及北部那个讲希腊语的城市生活、工作，度过了人生中一段重要的时光。我们并不知道，他是埃及本地人还是地中海世界其他地区的移民。任何对他生活年代的估计不过是猜测。丢番图引用过许普西科勒斯（约公元前150年）的一条定义，而赛翁（约公元350年）则引用过丢番图的一条结论。这就将他的生平限制在五百年的跨度之内，但更小的范围我们就没法知道了。

与其他希腊数学领域的作者留下的几何文本相比，这本《算术》极不寻常。它的主题不是几何，也不是日常计账的算术。它其实是一组复杂的问题，要求整数或分数必须满足特定条件。

例如，第二册的第八个问题要求读者“将一个正方形分为两个正
方形”。出于眼下的目的，我们可以将上述问题转变为更现代的
表述方式，从而能看出它与毕达哥拉斯三元组有关，即一个给定
的正方形（如前所示，记作c^2）可以分成或分解为两个较小的正
方形（$a^2 + b^2$）。当最大的那个正方形等于16的时候（这种情况 4
下答案涉及分数），丢番图展示了一种聪明的方式来求解；之后
他就转而去研究其他问题了。

然而，费马对此犹豫了，而且肯定问过自己这样显而易见的问题：这个方法可以扩展吗？人们能不能“将一个立方体分成两个”呢？这正是他在1657年向沃利斯和布龙克尔提出的问题（费马后来向他们通报说那是不可能的，随后沃利斯愤怒地反驳说这样的“负面”问题是荒谬的）。费马在页边空白处的提议实际上不仅适用于立方运算，也适用于任何更高次幂，这远远超出了丢番图所要求解的范围。

以上叙述中已经反复出现了另一个名字，所以让我们现在沿着历史的脚步从丢番图回溯到毕达哥拉斯，后者据信于公元前500年前后居住在希腊的萨摩斯岛上。尽管这一年代很早，但许多读者可能会觉得对毕达哥拉斯比对丢番图更加熟悉：作为数学史学家，我最常被问到的问题确实是“你会一路回溯到毕达哥拉斯吗？”毕达哥拉斯定理为人所知的确已有很长时间，令人失望的却是没有证据将其和毕达哥拉斯联系在一起。实际上，几乎没有证据将任何东西同毕达哥拉斯联系起来。若说丢番图是个身世神秘的人物，那么毕达哥拉斯便被神话与传说所笼罩。我们没有出自毕达哥拉斯或其直接追随者之手的文本。关于他的生平，有幸保留下来的最早记载出自公元3世纪，也就是在他

生活年代的大约八百年后，由作者们出于各自的哲学企图发掘而出。据说毕达哥拉斯在巴比伦或埃及学过几何；而这段出自推测的旅程，或许不过是那些作者虚构的，用来巩固他的地位与权威。至于他的追随者应该做过什么或者据信做了什么，这类故事可能有一定事实上的基础，但人们不可能确定其中的任何
5 一个。总之，毕达哥拉斯简直成了一个传奇人物。很多事都归在他的名下，可事实上人们对他并不了解。

毕达哥拉斯、丢番图、费马与怀尔斯，这四人的生活年代跨越了两千年的数学史。我们无疑可以追溯贯穿于每个故事中相似的数学思想，即使它们之间相隔了几个世纪。然后，我们就“完成”费马大定理由始至终的历史了吗？回答是“并没有”，而且原因还不少。第一个原因在于历史学家的一项任务便是将虚构从事实中剥离出来，并且让神话与历史脱钩。这并不是要低估小说或神话的价值：二者都体现了社会用来定义自身并理解自身的故事，而这些故事可能具有深刻而持久的价值。但是，历史学家一定不能让这些故事掩盖可能指向其他叙述的证据。在毕达哥拉斯这个例子中，比较容易看出貌似强有力的叙述是如何以及为什么会从最脆弱的话题中被编造出来；而就安德鲁·怀尔斯的例子来说，我们相信我们掌握了眼前的事实，也就更不易看出其中的问题。几乎所有故事的真相总是比我们最初想象的或是比作者有时想让我们相信的更为复杂，关于数学与数学家的故事也不例外。本章余下的部分便来探讨数学史上一些常见的神话和陷阱。为了方便起见，我将它们称作“‘象牙塔’版本的历史”、“‘垫脚石’版本的历史”以及“‘精英’版本的历史”。本书其余章节会接着给出另一些案例。

“象牙塔”版本的历史

怀尔斯的故事最显著的特征之一就是这样一个事实，即他自己故意与世隔绝长达七年之久，这样他就可以在不受打扰或干涉的情况下研究如何证明费马大定理。费马显然也是个孤独的人。如果没有别的原因，那就是地理上的距离将他和那些
或许已经能够理解并欣赏他工作的人隔开。我们已经谈到了丢 6
番图与毕达哥拉斯，却也没有提及他们同时代的人。难道这四位真是开拓新途径的孤独天才吗？这对数学研究来说是恰当的或最优的方式吗？让我们回到“毕达哥拉斯”这个主题接着谈下去。

与毕达哥拉斯相关的故事一直声称他在身边建立或吸引了一个团体或兄弟会。他们分享了某些宗教与哲学观念，也许还分享了一些数学上的探索。不幸的是，那些故事还声称兄弟会必须严格保密，这自然给人们对其活动进行无休止的猜测留下了空间。但即便这样的故事中只有一小部分是事实，毕达哥拉斯似乎也有足够的魅力吸引追随者。他的名字流传至今。这一事实的确表明他一生都受到尊重与敬仰，而且也表明他不是个隐士。

我们对确定丢番图的活动范围更有把握，他大概能在亚历山大享受与其他学者在一起的时光。几乎也可以肯定，他会有机会在庙宇或私人藏书中接触从地中海世界其他地区收集而来的图书。《算术》中的问题有可能是他自己的发现；但也有可能是他从比如书面文本或口述等其他各种来源汇编而成的一个单本合集。这本书重复出现的主旨之一，是数学经口口相传由一

个人传给另一人，并如此反复。像任何具有数学创造力的人一样，几乎可以肯定丢番图同某位老师或自己的学生讨论过他提出的问题和相应解答。因此，我们不应把他想成一个私下著书并沉默寡言的人，而应将他看作一位重视学习与知识交流的城市公民。

费马即便受限于图卢兹的区域范围以及全职政务的苛刻要
7 求，也并不像书中第一次出现时那样孤独。他早年在波尔多学习时有些朋友，其中一位叫艾蒂安·德·埃斯帕涅，此人的父亲是法国律师、数学家弗朗索瓦·韦达的朋友。韦达的作品原本并不多见，但由于这层关系，费马就有机会读到了。这注定会对他数学生涯的发展产生深远的影响。费马的另一位朋友是担任图卢兹顾问的同僚，叫皮埃尔·德·卡卡维。1636年，他移居巴黎时就随身带着关于费马及其发现的消息。费马经由卡卡维与马兰·梅森相熟；通过后者，他又与当时或许是巴黎顶尖数学家的罗贝瓦尔以及旅居荷兰的笛卡尔有了通信联系。后来，他同鲁昂的布莱士·帕斯卡及牛津的约翰·沃利斯交流了自己研究丢番图时的一些发现。因此，即便是远离重要研究中心的费马，也被连进了遍及欧洲的书信网络，这是一个虚拟的学者社区，后来被称为“文字共和国”。

说到怀尔斯，就更容易看到“孤独天才”故事中的裂痕：怀尔斯在牛津大学与剑桥大学接受教育，后来在哈佛、波恩、普林斯顿及巴黎等地从事数学研究，在所有这些地方，他都是蓬勃兴盛的数学社区中的一员。最终使他对费马大定理产生兴趣的数学线索源自他与同在普林斯顿共事的一位数学家偶然的谈话；五年后，他需要新突破时便参加了一次国际会议，来寻求该主题相关的最新思想；在证明的一个重要方面需要技术支持时他把秘密透露给了同事尼克·卡茨，并且把尚未解决的问题拿到了研究生讲座课中进行讨论，尽管最后除了卡茨外所有听众都走了；他在英国剑桥举行了三场讲座来公布整个证明，而在此之前的两周，他请同事巴里·梅热检查证明；最终的证明由另外六人检查；一个漏洞被发现后，怀尔斯邀请他以前的一位学生理查德·泰勒帮他一起进行修正。此外，在寻求证明定理的年月中， 8
怀尔斯从未停止教学活动或缺席院系研讨会。简而言之，尽管他一人独处了那么长时间，但也融入了一个允许他如此行事的社区，并在他需要时提供帮助。

怀尔斯独处的岁月吸引了人们的想象力，不是因为这对一位在职的数学家来说是正常的，而正因为这不寻常。数学从根本上来说必然是各个层次的社交活动。世界上每个数学院系都设有公共空间，无论是凹室还是公共休息室，而且总配有某类书写板，以便数学家们在喝完茶或咖啡之后聚在一起讨论问题。语言或历史学专业的学生很少合作撰写论文，也不会被鼓励这样做；但数学专业的学生经常富有成效地合作，而且还互教互学。尽管有现代技术的各种进步，但数学主要并不是从书本上学习，而仍然是通过讲座、研讨会及课堂来跟别人学习。

“垫脚石”版本的历史

在以上所勾勒出的费马大定理的故事梗概中，毕达哥拉斯、丢番图、费马及怀尔斯不仅在他们自己的生活中以孤独的形象出现，而且相互之间没有关联，就像一块块垫脚石立在一条原本就不起眼的河流上一样。如果说历史的象牙塔版本将数学家同他们的社团及社区分隔开来，那么垫脚石版本便将这些人与他们的过去分隔开来。由于“过去”被认为是历史的主题，因此以这种方式忽略其中很大一部分似乎很奇怪，但数量惊人的普通数学史却又是以垫脚石的形式呈现出来。

接着，让我们更仔细地重新审视我们的故事以及其中的历史空白。正如毕达哥拉斯和丢番图的人生轨迹鲜为人知一样，
9 人们对他们之间是否有所交往也不大清楚。丢番图可能从未听说过毕达哥拉斯。然而，几乎可以肯定他会接触到“毕达哥拉斯定理”，不过并非通过毕达哥拉斯的任何文本，而是从生活在公元前250年前后的欧几里得的著作中了解得知。除了这个非常粗略的年份以外，我们对欧几里得的了解并不比对几个世纪后的丢番图更多；但欧氏的主要著作《几何原本》流传了下来，成为有史以来使用时间最长的教科书，并且进入20世纪之后仍在学校的几何教学中使用。《几何原本》是欧几里得时代关于几何的一份综合汇编，其中的定理按照合理的逻辑顺序排列；而第一卷倒数第二个定理便是“毕达哥拉斯定理”，并借助几何构图被很好地证明。人们可以合理地假设亚历山大的丢番图有机会看到《几何原本》，并且有可能是“毕达哥拉斯定理”促使他对勾股三元组进行思考。不过，他的灵感同样有可能是来自我们已经无法了解清楚的其他来源。

即便是凭想象，丢番图与费马之间的最初几个世纪的情况也比丢番图之前的年代更难搞清楚。我们知道丢番图的《算术》最初成书时有十三卷，但只有前六卷以希腊语的形式保存了下来。至于方式及原因我们却不得而知。（1968年，在伊朗发现了一份阿拉伯语手稿，自名为第四卷至第七卷的译本。不过就该

译本对原文呈现的准确性，学者们还有争议。）幸运的是，拜占庭（后来的君士坦丁堡、现今的伊斯坦布尔）为希腊语世界保存了那六卷，而其抄本最后也被带到了西欧。正如第六章将进一步讨论的那样，1462年，一位名叫雷格蒙塔努斯的德意志学者在威尼斯见过其中一卷，而且他认为书中包含了欧洲人称之为“代数”的古怪主题的起源。一个世纪后，意大利工程师、代数学家拉斐尔·邦贝利在梵蒂冈研究了《算术》的手稿，接着他把写作自己的代数书的工作停了下来，转而去整理丢番图的问题。第 10
一个印刷本于1575年在巴塞尔出版。这是一个拉丁文译本，由威廉·霍尔茨曼翻译、编辑。霍尔茨曼是一位人文主义学者，他对作品的描述是“无与伦比，并包含着算术真正的完美”。丢番图的问题持续引发那些与之接触的人的兴趣。1621年，克洛德·加斯帕尔·巴谢·德·梅齐里亚在巴黎完成了新版的拉丁文《算术》。费马所拥有并在上面做记录的，正是这个版本。

填补费马与怀尔斯之间的历史空白倒不是太困难。塞缪尔·费马于1670年公开的费马大定理，在17世纪似乎并没有吸引什么人来认真地尝试证明。但在18世纪，它引起了莱昂哈德·欧拉的关注，欧拉是那一时期最有才能且最多产的数学家，他介入了其中一些比较简单的例子。1816年，巴黎科学院为征求该定理的证明设立了一个奖项，这激发了索菲·热尔曼的努力尝试，她在其中某些方面取得了一些进展，而她的工作又由另一些人接手并拓展了下去。此外，这个问题慢慢也广为人知，许多年间吸引了就算没有成千上万也有成百上千人投稿。他们中既有专业人士也有业余爱好者，都声称已经完成了对该定理的证明。其中的大多数尝试都是错误而无用的，但有些尝试本身则推动了重要的数学发现，怀尔斯对此应该有所了解。当他终于着手自己证明的时候，他运用了一些20世纪最深奥的数学原理，也就是至那时为止已知的与费马大定理有关的成果：由两位日本数学家在20世纪50年代提出的谷山–志村猜想，以及由俄国的维克托·科里瓦金与德国的马提亚斯·弗拉赫在20世纪80年代提出的科里瓦金–弗拉赫方法。我们再次注意到，数学家的习惯是将他们前辈的姓名写进历史。我们同时还注意到，单个定理的背后是历史相互影响所构成的复杂网络。

一般说来，越往前回顾，追溯各个“垫脚石”之间的历史情
11 况就越困难，这还不单单是因为许多证据早就被冲走了。但若不尝试，就没有历史，而只有一系列逸事，许多流行的数学史仍然常常以此为基础。

“精英”版本的历史

尽管我们对欧几里得或丢番图的生活几乎一无所知，但还是有几件我们可以肯定的事情：二者都受过良好的教育，并且能自如地用希腊语写作，这是当时地中海东部知识阶层的语言；他们都有机会接触早期的数学著作；两人都能理解、整理并扩展当时的一些前沿数学；他们所涉及的数学没有实用价值，而只是纯粹的智力游戏。即使在像亚历山大这样的城市，从事此类数学工作的人士的数量也从来不可能有多高。实际上，据估计，在通行希腊语的地方任何时候都只有少数几位这样的人士。换句话说，欧几里得与丢番图都属于极少数的数学精英。

只需片刻的回顾便足以说明，与他们涉足的领域相比，数学肯定有更广泛的用处。希腊社会与其他社会一样，有店主、

管家、农民及建筑商，还有其他许多人士要从事日常测量与计算。我们对他们的方法几乎一无所知，因为这些人大多会通过示范样板及口口相传的方式来进行学习与教学。他们也没有被组织成学校或行会，尽管我们确实知道一个有名的团体，即“harpēdonaptai”，或者不妨称之为“拉绳定界师”。他们所用的数学自然很少留下痕迹。成堆的标记，或者在木头、石头或沙地上刮刻的记号，一旦没用就会立即被丢弃，而且肯定不会存放在图书馆里。不管怎么说，都是社会地位相对较低的人从事这些行当，而身处学术圈的知识分子对此却少有甚至没有兴趣。 12

当数学史学家说起“希腊数学”时，就像他们经常做的那样，几乎总是在谈欧几里得、阿基米德、丢番图等人留给我们的复杂的书面文本，而不提及普通民众用于日常或园林的数学。但最近，这种情况已经有所改变。历史学家开始承认，精英式的希腊数学起源于地中海东部日常的实用数学，虽然后来的数学家通过发展一种更为正式而“无用”的数学来同那些起源保持距离。

对于“希腊数学”这一笼统的表述还有些要注意的地方。当时，丢番图住在埃及的亚历山大；阿基米德住在西西里岛的叙拉古；而另一位伟大的“希腊”数学家阿波罗尼乌斯则居住在佩尔加（所在区域现属土耳其）；换句话说，尽管他们都用希腊语写作，但都不是来自我们如今称之为希腊的地方。的确，就我们所知，丢番图可能是非洲人生养的。不过，文艺复兴时期欧洲人高度推崇的“希腊数学”已经被认为本质上是“欧洲的”了。当我们想到大陆另一端的西班牙被排除在欧洲之外时，将亚历山大纳入欧洲的荒谬性就变得更加明显。公元8世纪初，西班牙处在伊斯兰统治之下，因此享有伊斯兰世界丰富的文化与知识。然而，人们经常读到，阿拉伯数字是由13世纪初在意大利比萨进行写作的斐波那契引入欧洲的，好像在此之前阿拉伯数字在西班牙使用了两个世纪之久算不上什么，而且好像西班牙在某种程度上不是欧洲的组成部分。那些促进精英数学事业发展的人，自然倾向于将任何会赋予他们的学科以权威及受人尊敬之地位的内容吸收进其历史之中，而不考虑其他会招致麻烦的事实。

无论数学应用在哪里，我们都有可能找到一些高等级且备
13 受尊敬的从业者，但更多的人名将永远不会被载入任何史册。假如我们重新审视费马时代的情况，那么会发现几乎没什么不同。在他的一生之中，法国的精英式数学活动异常丰富：人们可以想到在巴黎有那么三四位与费马保持联系的人士。粗略地估计，在荷兰和意大利或许也有那么几位，甚至在英国还有一两位，但也就仅此而已。然而，社会等级较低的数学研究活动要比人们所预期的更为普遍。近期检索数字化资料的结果显示，在16世纪与17世纪英国出版的书籍中，有多达四分之一的书籍以这样或那样的方式提到了数学，即使只是顺便提到。此外，面向想获得基本数学技能的商人或手工业者的书籍在稳步增多。

在结束本章之前，就让我们更详细一点地来看看其中的一本：毕竟，没有比考察原始资料更好的方法来探索数学史了。罗伯特·雷科德的《通往知识之路》于1551年在英国出版；费马就在大约五十年后出生。雷科德一生中的大部分时间都是医生。1549年，他被任命为布里斯托尔造币厂的总监，两年后又被任命为爱尔兰银矿的检验官。不幸的是，在此期间他招惹了

政敌，最终身陷伦敦的国王监狱，并于1558年在那里去世，享年四十八岁。不过，雷科德大多数的数学著作也正是在这段时间里出版的；如今他因这些著作而被世人所铭记。雷科德在牛津大学和剑桥大学接受教育，能说流利的拉丁语和希腊语，但他做出了大胆的选择，用英语来写数学方面的文章。特别是，他旨在让普通民众也能够使用作为最精英的数学家之一的欧几里得的数学。这并不是件容易的事：一方面，尽管大多数英国工人或许已能熟练地掌握铅垂线和直尺，但他们却从未听说过一个被称作“几何”的正式科目；另一方面，当时英语里根本也没有词汇来表达诸如“平行四边形”或“扇形”之类的专业术语。雷科德 14
凭借想象力和技巧解决了这两个问题。

在冗长的序言中，他从最低下的社会地位开始一路往上，描述了几何对哪些阶层的人来说是“非常必要”的。处于最底层的是在土地上耕作的“无知者”。雷科德解释道，即使是这些人，对几何也有出于本能的掌握。否则他们的沟渠就会坍塌，他们的干草堆也会倾倒。再向上到了生意人阶层，雷科德反倒提供了一份长长的清单，列出了几何对哪些人是不可或缺的，如商人、海员、木匠、雕刻师、细木工、泥瓦匠、画工、裁缝、鞋匠、织布工等，更进一步得出了如下的结论，

> 即手工技艺从来都没有像出彩的几何那样富有机智，且对世人来说又是如此必需。

雷科德尤其认为几何在医学、神学及法律专业中也是必不可少的，尽管他的论点随着其社会地位的攀升变得更加做作且

缺乏说服力。

雷科德很能理解普通人的难处，这在他谈到几何本身时表现得最为明显：他的论述是一种良好的教学法模型，用通俗易懂的语言来表达，并配有许多示例和有用的图表。他很早就教授过如何用欧几里得定义的直尺与圆规构造直角。不过，万一发现这种方法过于困难，那么他还有另一项可供选择的建议：画一条直线，再分别画出3个、4个和5个单位长度，然后用这些长度来构造一个三角形。其短边之间的夹角就是直角。这并非经典的欧几里得式构图方式，对于实际操作者及“拉绳定界师”来
15 说，却是一种实用的方法。

对于21世纪，我们可以做一份比雷科德的更长的清单，来罗列那些日常生活里比如在学校、居家或工作场所中使用数学的人士。我想起了我的母亲艾琳，她在八十九岁时既不相信银行也不相信计算机，而是在精心划分好区隔的笔记本上算着她用于家庭开支的每一个便士。我也想起了我的好友塔季扬娜，她反复跟我说她在学校时数学并不好，但她却创作了设计复杂的拼图（见图1）。她当然可以处理直角三角形。甚至，她对镶嵌式布局及比例的直觉或许让她有资格成为“拉绳定界师”这
16 个群体的现代代表。

“精英”版本的历史中并没有艾琳或塔季扬娜的一席之地：尤其是女性，在被认真对待之前，至少得先将自己提高到索菲·热尔曼的水平。但如果没有在各个层次上从事及教授数学的人，精英人士就无法蓬勃发展。在怀尔斯、费马或丢番图所占据的先驱领域背后，延展着数学活动的广阔腹地，这在一般的数学史中还只有很少的讨论。本书的部分目的就是要恢复这种平

图1 “色块画作”，塔季扬娜·特克尔·佩佩作。她自称并不擅长数学

衡，为包括孩童在内的所有人找回数学，并从一些新的角度重新审视数学的历史。17

第二章

数学是什么及数学家是什么人?

在上一章中,我设想读者会将“数学”或多或少地看作他们在学校时学习过的那门如此命名的学科,并且把“数学家”视为成年后继续从事这门学科研究的那些人。不过,历史要求我们更仔细地考察这两个概念。我个人的体验是这样:当我还是学校老师的时候,发现自己整个上午都在教授百分比、关于圆的定理以及微分学的课程。我被迫问自己,这种内容上令人难以置信的汇集是如何统一在“数学”这个单一主题之下的。大多数人或许都会同意这样一个笼统的说法,即数学是以空间与数字的性质为基础。但他们又是怎样看待流行的数独游戏呢?这是否是一种数学性质的消遣?我听过专业的数学家就此争论不休,而无论是赞成抑或反对,他们争论得都很激烈。

让我们回到起点。希腊语“mathemata”一词的本义只是“所学到的”,有时是常规的意思,有时更确切地是与天文学、算术或音乐有关。这个希腊语词汇构成了现代英语单词“mathematics”及其他欧洲语言中其同源词的词源(例如法语中的“mathématiques”、德语中的“Mathematik”、意大利语中的“matematica”，或者美国英语中的“math”）。不过，“mathematics”一词的意思，在数 18
个世纪间经由许多变化而出现了偏差和扭曲，这些我们不久就会看到。但这只是从欧洲的视角看待问题。如果我们回溯一千年或两千年，在欧洲文化成为主流之前，我们能否找到与“mathematics”一词对等的汉语、泰米尔语、玛雅语或阿拉伯语词汇？如果能，那么这些词汇又涵盖了哪些著作及活动呢？要想彻底研究这个问题，对一群学者来说将是毕生的工作。但就像在本书其他地方一样，本章中一些案例的研究将用来阐明需要提出的问题以及可能会出现的答案。

追溯“算”字的含义

从中国官修史书对公元前200年之前不久至公元200年（秦汉时期）的记载中，有可能发现二十多人的名字，据说他们在“算”的某些方面已经很熟练。“算”作名词时可以表示一组用木头、金属或象牙制成的短棒，人们在平面上熟练地操作这些短棒来标记计算中的数字。“算”一字也可以表示用短棒进行计算的行为。这便是数学活动的证据，但我们对此还是不太了解，除非我们能发现进行了哪种计算。

就官方记录中提到的许多从业者来说，“算”似乎与被称为“历”的天文或日历系统密切相关。现代社会之前的所有社会都使用太阳、月亮及行星的位置，来为宗教仪式或农作物种植确定适当的时间和日期。所以，那些能够根据天文数据做出正确预测的人对于统治者及政府来说是必不可少的。因此，在早期古代中国的历史中，“算”与“历”时常有着联系。然而，相同的

记录也表明，“算”与更多的世俗事务、利润计算以及资源分配
19 有关。

20世纪80年代初人们发现了一份新的原始资料，大约来自公元前200年这一时期，可以进一步说明“算”在当时的使用情况。这份名为《算数书》的文本刻在了一百九十根竹片上。这些竹片每根约三十厘米长，起初是用结绳并排地连在一起，所以可以像垫子一样卷起来。最后的“书”字意为“写就的文本”，有时也指书籍的“书”。中间的“数”字可以概括地解释为“数目”。不过就我们的目的来说，最相关的是“算数”这个词的含义。《算数书》的文本中有约七十个问题，并附带解题的说明。这些问题包括：整数与分数相乘，按照不同贡献者所投入的金额来分享利润，将商品生产中的浪费考虑在内，从给定数量的价格计算总成本，税收的计算，找出混合物中不同成分的数量，将一定数量的原材料转化为若干成品，检查旅途所花的时间，体积与面积的计算以及单位换算等。

因此，《算数书》中的问题很大程度上源自日常活动及交易。这本书的写法非常直接：作者对每个问题设“问”，再给出“答案”以及“解题方法”。以下的示例是第二章中的两个“通关问题”：

> 一只狐狸、一只野猫和一条狗经过关口时，被征收了一百一十一钱的关税。狗对野猫说：“你的毛皮的价值是我的两倍，你应该支付两倍的税金！”而野猫对狐狸也这么说。问：每个角色分别支付多少税金？答案是：狗支付了十五又七分之六钱，野猫付了三十一又七分之五钱，而狐狸

付了六十三又七分之三钱。方法如下：让狐狸所付的税金是野猫的两倍，而野猫支付的又是狗的两倍。这样相加得七个单位，从而有了除数。再用占比分别乘以税金就得到了被除数。这样每次用被除数除以除数就得到一个解。 20

而更实际的或许是这样一例：

一个人带着脱壳的谷物过三个关口。我们并不清楚有多少谷物。（每个）关口征收三分之一的税。离开关口后，他还剩下一斗的脱壳谷物。问：他动身时带着多少脱壳的谷物？答案是：他带的脱壳谷物有三斗三升又四分之三升。方法如下：从一斗脱壳谷物开始，经三次翻倍就得到了除数。再从一斗开始，先三倍，再三倍，接着乘以通关次数即可获得被除数。

这些答案都正确，但对“方法”的描述不大好懂，很可能是需要通过口头解释来加以补充。说明也仅是对所述问题中的特定数字而给出的，不过训练有素的读者可以将其应用于任何类似的问题。所以从这个意义上讲，那些说明提供了一种通用的技术。然而，这本书没有期望读者去理解方法背后的推理过程，而只是希望读者能应用方法。

类似的以及其他的问题出现在一份稍晚的文本中，即《九章算术》。这是分九个部分来讨论“算术”的著作，因此通常被称为《九章》。正史表明，这份文本在公元2世纪初就得到使用了。不过，与大约三四个世纪之前欧几里得的《几何原本》一样，我们既没有与《九章》作者或编写相关的信息，也没有原始文本。我们今天唯一能看到的是刘徽在公元263年编撰的版本。在《算数书》的内容于2000年得以誊抄并出版之前，《九章》一直是最早的专门研究“算”的广泛文本。因此，《算数书》的发现不仅使重要的文本比较成为可能，而且还为历史学家提供了一份更
21 深入的信息，来了解早期古代中国“算”的使用情况。

即使从以上的简述中也可以清楚地看到，“算”字与我们用“数学”（mathematics）一词所能得到的任何重要主题无关。相反，该字是指可以在多种情况下使用的技术与技巧，从日常应用到朝廷所需的天文演算的“历”，再到更平常的“算数”。现在转向拉丁西方，我们能找到与“数学”一词相关的一系列类似的实践活动吗？

追溯“数学”一词的含义

大约在公元100年，罗马作家尼科马库斯列出了涉及数量多少与规模大小的四个学科：算术、音乐、几何以及天文学。对他来说，算术是关于数量多少（或数字）的研究，而几何是对规模大小的研究，二者是最基本的；音乐是关于彼此间数量多少的科学，天文学则处理运动中的规模大小。四个世纪后，哲学家波伊提乌将这些学科统称为“四艺”。它们连同语法、逻辑和修辞学这“三艺”一道，构成了中世纪学术课程的七门人文学科。波伊提乌自己则撰写了算术及音乐方面的专著，整个中世纪欧洲的大学都在研究这些论述。一些关于几何的著作也归在他的名下，但真正的原著者却并不确定：波伊提乌和毕达哥拉斯一样成了一个神话般的人物，而以后的工作都可以顺当地归功于他。

算术与几何至今仍处于数学的核心（我们或许还记得艾琳和塔季扬娜也曾实践过），而天文学和音乐现在已走上各自的发展道路。分道扬镳发生在17世纪。当时，在数学理论与音乐实践之间进行调和变得越发困难；而天文学则努力摆脱与占星术的长期关联，依托自身成了受人尊崇的学科。 22

无论如何，到了文艺复兴时期，尼科马库斯划分的“四艺”范围开始显得过于狭窄，无法涵盖随着财富、商贸及旅行的迅速增长而出现的众多新的数学活动。约翰·迪伊在1570年的首个英译版欧几里得《几何原本》的序言中列出了一份数学门类与科学的“基础框架”或蓝图（见图2）。算术与几何仍然是其中关键的组成部分；不过到目前为止，用来回答“多远”、“多高或多深”以及“多宽”的几何学已经产生了“地理学”、“生物地理学”、“水文学”以及一个名为“战略数量学”的门类。此外还有一份长长的名录，列出了被认为是算术与几何的“衍生”学科，其中就包括“天文学”和“音乐”。现代的读者对“透视”、“宇宙结构学”、“占星术”、“静力学”、“建筑学”及“航海”等词汇的含义还有所了解，但可能会对“人类学”、“气动学”、“科学观察学”以及其他几个不常见的门类感到困惑，就像那时的读者对此或许已经感到了困惑那样。确实，晦涩难懂的主题以及小标题和子标题下的整齐划分表明，迪伊的系统化就像尼科马库斯或波伊提乌更为简单的方案一样，是一种哲学实践，而不是对现有实践的真正分类。

那么，我们如何才能更精确地发现公元500年至1500年这数世纪间西欧所进行的数学活动呢？我们能否像对“算”字那样来对“数学”进行相同的研究，通过检查使用该词的语境来发 23

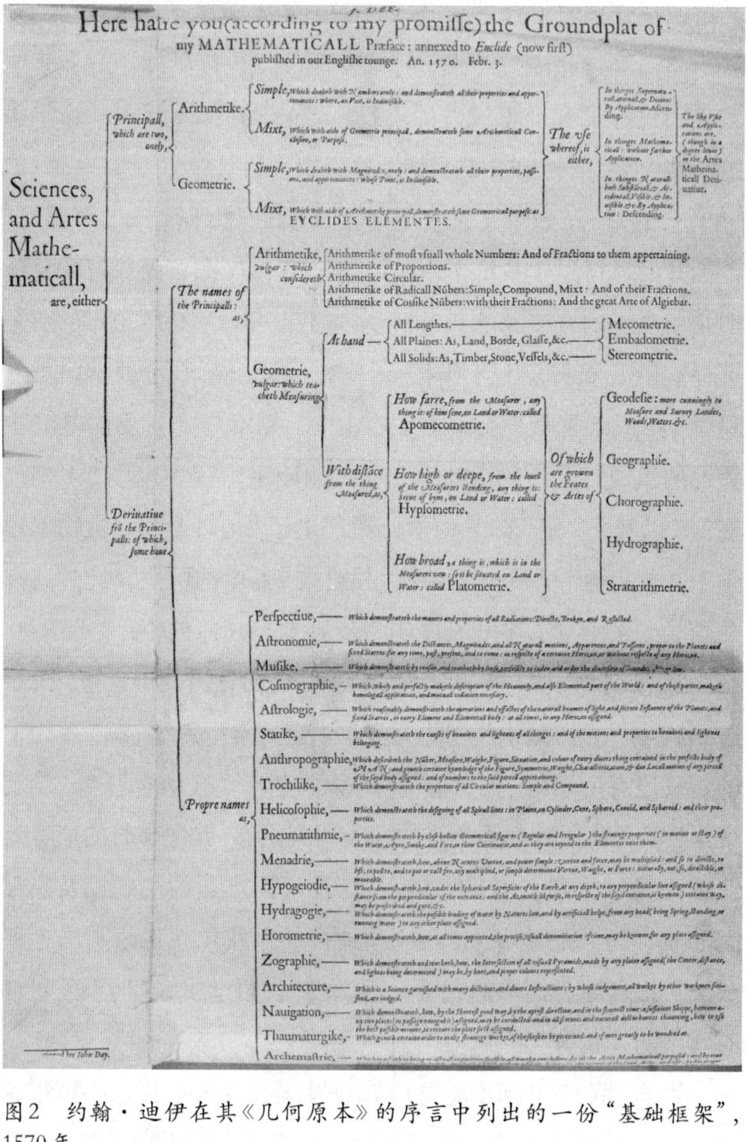

图2　约翰·迪伊在其《几何原本》的序言中列出的一份“基础框架”，1570年

现其含义呢？这一时期留存于西欧的文本远远多于早期古代中国的，所以不可能进行全面的考查。但作为首次尝试，特别是由于其与英格兰著作者有关，我们将研究由荷兰学者赫拉德·约翰·福西厄斯编写的一部数学史，也就是他的《数学科学》。该书于1649年在阿姆斯特丹出版。 24

向荷兰学者找寻涉及英国思想史的信息，这看起来或许很奇怪，但福西厄斯对英格兰著作者的大部分叙述是基于英格兰文物研究者约翰·利兰的早期工作。1533年，也就是在解散修道院之前的几年，利兰受亨利八世委托搜寻其治下的图书馆与学府，并列出其中的收藏。在接下来的两三年中，他列出了约一百四十个宗教机构的财产。随后书籍的流失让他非常痛心，1536年，他向托马斯·克伦威尔抱怨道："日耳曼人察觉到我们的懒散和疏忽，便日复一日地派遣年轻学者前来，他们破坏了馆藏书籍并把它们带出了图书馆。"利兰提供了最后的也是最全面的图书馆馆藏记录。他打算编纂一本英格兰著作者词典，这本词典包含六百多个条目。但不幸的是，在词典完成之前他就精神失常了。不过，他卓越的工作得到了其他历史学家的认可，后来包括福西厄斯在内的许多作家都直接或间接地借鉴了他的发现。

福西厄斯提到的最早的英格兰作家是贝德，他对公元730年前后这一时期进行了记述，被列在"天文学"和"算术"的类别下。贝德在英格兰东北部贾罗的修道院里度过了他一生中的大部分时光，他以评点《圣经》以及在教会中研究历史而闻名，但现在很少有人把他看作天文学家。然而，归在他名下的那些著作却叙述了在月球之上及月球的周期、复活节的日期、行星和黄道十二宫、星盘的使用以及春分的计算。其中一些著作可能被后来的评论者错误地归在了贝德的名下；不过，贝德显然非常关心复活节在哪一天，这个日子对基督徒来说至关重要，就像准确的冬至日期对于古代中国皇帝的那种重要性一样。这个计算同样不容易：复活节必定是春分之后的一轮满月后的第一个星期日，因此要正确计算该日期就需要了解月球与太阳的周期，而
25 两者之间并不自然相关。在英格兰北部、爱尔兰及罗马存在着两种基督教传统，这导致日期上的冲突，而这种情况最终于公元664年在惠特比会议上得以解决。贝德本人可能没有进行必要的计算，但他知道有什么风险。

最终，与教会计时相关的运算以“计算”之名而为人所知，并且在整个中世纪时期仍然至关重要。

继贝德与他的追随者阿尔库因之后，四个多世纪中再也没有英格兰人的名字出现在福西厄斯的记述中，直到我们遇见1130年前后来自巴斯的阿德拉德。阿德拉德似乎曾经在法兰西、西西里岛及叙利亚旅行。他是最早将欧几里得《几何原本》的若干部分从阿拉伯语译成拉丁文的译者之一，而且据说他也曾在星盘上写作。

仅仅在13世纪和14世纪，诸多人名（及推定的年代）才开始以越来越高的频率出现，且全都归在“天文学”或“占星术”的名下：约翰·萨克罗博斯科（1230），他关于地球及它在宇宙中地位的著作在长达四个世纪的时间里一直是大学课程的重要组成部分；罗杰·培根（1255），被描述为一位占星学家；沃尔特·奥丁顿（1280），据说写过关于行星运动的文章；来自北安普敦的罗伯特·霍尔科特（1340），据说写过星体运动；约

翰·伊斯特伍德（1347），占星学家；尼古拉斯·林恩（1355），占星学家；约翰·基林沃斯（1360），天文学家；西蒙·布雷登（1386），据说写过关于医学、占星术及天文学的文章；约翰·萨默（1390），占星学家；等等。接着在15世纪，这些名字又开始淡出历史。显然，14世纪是天文与占星研究的高峰，一个促成因素可能是1348年“黑死病”所造成的可怕打击。上述那些人中有许多都属于某个宗教团体，比如（天主教）方济各会、多明我会或加尔默罗会。许多人还与牛津大学特别是同墨顿学院有关联，而直到今天他们的一些著作仍被安全地保存在牛津大学图书馆。他们都一而再，再而三地跨越了天文学与占星术之间游移不定的界限。 26

与这份天文学家名录不同，没有任何英格兰著作者出现在福西厄斯有关音乐、光学、大地测量学、宇宙学、年代学或力学的章节中，只有蒂尔伯里的杰尔斯和罗杰·培根作为地图制作师在地理学章节被提及。因此，从16世纪的视角回顾中世纪英格兰的数学著作，占主导地位的主题是计算与占星术。

然而对于欧洲其他地区，情况或许会有所不同。比如在位于地中海西部心脏地带的意大利，其贸易情况比在北欧更加广泛和复杂。13世纪时，建成了算盘学校来训练男孩子们学习商用算术，甚至还有些初步代数（求解某些基本方程）。开创性的文本则是比萨的莱昂纳多所著的《计算之书》，而他后来也以“斐波那契”闻名于世。《计算之书》里有数百道营商方面的问题。其中两例如下：

> 四名男子合伙经营，其中第一个人投资了总额的三分之

> 一，第二个人投资了四分之一，第三个人投资了五分之一，第四个人真就投资了六分之一，他们的总利润为六十索尔第；求每人持有多少。这个问题的确与下述题例相同，即四个人以六十索尔第的价格购买一头猪，其中第一个人希望拥有这头猪的三分之一，第二个人希望拥有四分之一，第三个人希望拥有五分之一，而第四个人希望拥有六分之一……

莱昂纳多本人已经点出了这一问题的两个版本。从数学上讲，它也等同于《算数书》中那道狐狸、狗与野猫通关的问题。下一个问题则反映了当时意大利所关心的事情，并且是数百种涉及货币兑换或材料转换问题中的典型题例。同时，它表明在丢番图之后约十个世纪，另一种算术仍在亚历山大城蓬勃发展。

> 在亚历山大，十一卷热那亚布料值十七克拉；那么九卷佛罗伦萨布料值多少钱呢？因为十一卷和九卷不是同一个重量数值，所以你得用十一卷热那亚布料来求佛罗伦萨布料的价值，或者用九卷佛罗伦萨布料来求热那亚布料的价值，以便两者要么是佛罗伦萨的，要么是热那亚的。但由于你很容易就能求得佛罗伦萨布料，而每一卷热那亚布料对应二又六分之一卷佛罗伦萨布料，那你会用这个比率乘以热那亚布料的卷数，就得到二十三又六分之五卷佛罗伦萨布料……

27 对于他们的学问来说，福西厄斯和他在北欧的信息来源从未见过《计算之书》；福西厄斯只是通过传闻知道有这么一本书，而且还把书的年代弄错了两个世纪。由此可见，当时数学领域的活动可能在很大程度上受到地域限制。

这也跟时间有关。中世纪时期，大多数后来由迪伊和福西厄斯所发明的主题类别在很大程度上都已是多余的，至少对英格兰来说是这样。至16世纪末，随着整个不列颠都进入更广阔的世界，情况便不再如此。在1600年前后从事研究的托马斯·哈里奥特留下了涉及光学、弹道学、炼金术、代数、几何、航海与天文学等领域的著作。在此期间，与他同时代的西蒙·史蒂文在荷兰发表了类似的文章，只是航海由（对他来说）更相关的锁具和水闸问题所代替。计算与占星术已经让位于一种新兴世界秩序下的数学活动。

数学是什么？

如果确实有数学这样的存在，那么历史上的数学又是什么呢？现在应该很清楚，数学活动已经有了多种形式，但它们只是由于需要某种度量或计算的事实而被松散地联系在一起。一个更精确的答案在很大程度上必然取决于时间和地点。目前有一些共同的脉络：所有组织有序的社会都需要规范贸易活动并校准计时，这大致说来分别是早期古代中国算数与算历或者是13世纪的欧洲算盘和计算的目标。然而，各种技术的实践者们可能有着非常不同的社会地位。算数与算盘的教学是面向商人或官员的，而算历或计算则是中国高等级专门人士以及中世纪欧洲僧侣和学者的领域。“高级的”数学通常需要一定程度的抽象 29
思维。数百年间，不同背景下，在受过充分教育而从事这种数学的人士和使用“一般”或“市井”数学的商人或匠人之间，反复出现了地位与受尊重程度的分离。

随着社会变得越来越复杂，对数学的要求也越来越高。迪伊所列出的一长串科目，即便有些多余，也表明有大量活动都运用了数学的专业知识。这些主题统称为“混合数学”，这表明数学是各个科目不可或缺的一部分（这与后来“应用数学”的概念还不大一样，后者使用数学来分析自身以外的主题）。

没有理由认为，从早期古代中国及中世纪欧洲所吸取的教训不会扩展到其他社会：没有某种单一的知识体系可以让我们方便地称之为“数学”，但我们可以确定许多数学学科和数学活动；而哪些特定的学科和活动被认为是最有意义或最负盛名的，则总是时间和地点方面的问题。

数学家是什么人？

既然我们已经开始确定构成数学的那些活动的范围，那么我们可以说谁算或不算数学家吗？毕达哥拉斯、丢番图、费马和怀尔斯，这四位通常都被描述为数学家。前三位虽已作古，他们的姓名却被收入一套标准的参考书《数学家传记词典》。然而，这三位中没有谁会认出赋予他们的这项标签。我们根本不知道
29 毕达哥拉斯会如何描述自己。丢番图或许会认为自己是位算术学家，即并非算术或算盘这类日常运算而是“高级算术”的实践者，后者研究的是自然数某些较为晦涩与难于理解的特性。另一方面，费马可能会自称为几何学家，当时几何是四艺中最具权威且最受人尊敬的分支。这在法国直到19世纪仍然是对致力于学术的数学家的标准描述。这四人当中，我认为，只有怀尔斯会毫无保留地自称是数学家。

如今，数学学科受到高度尊重，甚至备受尊崇。但从本章已经谈到的内容可以很容易地看出为什么情况并非总是如此。12世纪，索尔兹伯里的约翰声称，数学的实践，即由恒星与行星的位置预言未来，源自世人与恶魔之间的宿命纠葛，就像手相术（解读掌纹）那样；而占卜（解释鸟类的飞行）则是邪恶的源头。1570年，医生及文艺复兴时期最重要的代数学著作之一的作者吉罗拉莫·卡尔达诺因推测耶稣基督的星象运势而入狱。1605年，托马斯·哈里奥特因与“火药阴谋”实施者有联络这一罪名而被捕。相比于案情本身，他更多是因为把詹姆斯一世的星象运势钉在自家墙上这件事而受审。而在17世纪末，约翰·奥布里写到那位乡村牧师、数学教师威廉·奥特雷德时说：“乡下人都确信他会召唤魔法。”在前现代的欧洲，“数学”的实践对实践者及其设想的科目来说并非没有危险。

事实上，从1570年起“数学家”一词才开始在英语的数学著作中频繁出现。最初，这个词汇主要用在外国作者身上；但后来很奇怪地在两种毫不相干的语境中用于枪炮手或占星学家。1660年王朝复辟后，该词逐渐被更广泛地用于算术或几何学的作者，不过对占星学家仍然适用。同时，“数学”的预言成为讽刺 30
与嘲笑的常规主题。数学与占星术长期且持久的联系有助于解释为什么学者们宁愿避免使用该术语。当亨利·萨维尔于1619年在牛津大学设立两个数学教席时，它们分属几何学和天文学，且有着严格的指导方针，即后者不应包括国运占星术。时至今日，剑桥大学仍然有一位卢卡斯数学教授，而在牛津大学地位相当的则是萨维尔几何学教授。而且，除非我们认为数学同预言及其影响的关联只是一个发生在欧洲的现象，否则值得记住的是，现代汉语中代表“mathematics”一词的术语“数学”，传统上是指占卜背景下的数字研究。

简而言之，正如我们现在所理解的那样，“数学家”一词是现代欧洲的发明。在数学活动的漫长历史中，有数学家的时间非常短。而假如我们想恰当地欣赏数学史，重要的就是不要将他们的形象投射回历史。因此，历史学家倾向于使用更精确的描述，例如“抄录员”、“宇宙学家”或“代数学家”，或者像“数学从业者”这样更通用的表述。有件事可以肯定：数学的历史不

31 是数学家的历史。

第三章

数学思想是如何传播的?

上一章中我们对不同时间和地点的数学活动进行了广泛的调查。这是研究数学史的一种方法,即确定人们的实际行为。然而,历史学家总是想问更多的问题:不光是人们曾经知道什么,还有他们彼此之间以及与后人之间是如何进行交流的。数学思想如何从一个人传给另一个人,怎样从一种文化传播到另一种文化,或是如何从一代传承给下一代?(回想一下在第一章中首先提出的那些问题:费马是如何知道丢番图,或者怀尔斯又是如何知道费马的?)

这些问题延伸下去,便是要问历史学家自己是如何了解历史中的数学的:我们有哪些原始资料,这些资料是怎么流传到我们手中的,它们的可信度如何,而我们可以怎么来学着阅读这些资料?本章将探讨数学思想在某些时候跨越漫长时空的方式又为何在另一些时候无法做到这一点。

脆弱、稀缺且晦涩的原始资料

早在一千多年前,埃及和现今伊拉克地域内的人们就已经

32 进行了复杂的数学实践。那些自信数学是始于毕达哥拉斯的人，现在可能会因为这个发现而感到轻微的眩晕。公元前两千纪和公元前一千纪的埃及文明和巴比伦文明彼此间相对接近，不过，我们对后者数学知识的了解要多过对前者的。原因很简单，底格里斯河与幼发拉底河沿岸用作书写材料的黏土板坚固耐用，而尼罗河地区用纸莎草制成的纸张却不然。人们从伊拉克挖掘出了数以千计的黏土板，其中许多带有数学方面的内容。或许还有数以千计的黏土板仍然埋在地下，假如它们没被坦克履带压碎或者没有在最近的战乱后被洗劫一空的话。另一方面，就埃及来说，用三只手的手指就可以数出尚存于今的数学文本与碎片的数量，而且它们散布在一千多年的历史长河中。英国的情况也一样，或许仅存“诺曼征服”前后以及19世纪时期的些许文本。显然，现今尚存的埃及文献仅仅提供了微乎其微的深入探究空间，但同时却为埃及数学活动的相关推测和幻想留下了充分的空间。

在印度、东南亚和南美，情况与在埃及大致相同：气候迅速破坏了诸如木材、皮肤或骨头之类取自天然的书写材料，以至于历史学家不得不竭尽所能，从为数不多且保存也不好的文本中了解信息。显然，文献的匮乏使我们对历史的印象失真了。我们必须确认，遗存至今的物品是否代表了已经逝去的东西。要知道，一个新发现（比如中国的《算数书》）就可以从根本上改变我们对整个数学文化的认识。不过，文本的缺失或许有些好处，因为这迫使历史学家加大对原始资料的搜集。例如，行政记
33 录可以揭示日常生活中进行的计数与测算。考古证据加强了我们对如何规划并建造建筑物，为此又必须进行哪些计算的了解

（因为我们没有直接证据表明，巨石阵或金字塔的建造过程涉及计算）。图片、故事或诗歌等各种原始资料也可能包含同一时代数学知识的线索。

许多古老的文本是用现今已不复存在的字母和语言写成的，翻译这些文本过程充满困难。拥有必要的语言技能并勇于接触数学材料的学者确实依然很少；他们的任务却极其微妙。任何将一种语言转成另一种语言的翻译都可能破坏源语言的实质，而数学翻译带来了进一步的困难，即如何让现代观众理解另一种文化的技术概念。比如下面一段文字，摘自公元628年印度的《婆罗摩历算书》，普通读者对此可以得出些什么呢：

> 山高乘以给定的乘数便是至城市的距离；这个距离不会被擦除。用所得距离除以增加两个单位后的新乘数，便是其中一人跳起的高度。前提是两人完成了一样的行程。

为理解这个问题，读者需要知道，一个旅行者从山上下来，沿着平原步行到城市；而另一个旅行者则不可思议地从山顶跳到更高的垂直高度，然后沿着斜边飞行相同的距离。对当时的学生来说，这或许是标准问题之一（该问题的另一个版本是猴子上树），并且有可能通过口头解释加以阐明。但对21世纪的读者来说，不懂梵语或7世纪印度数学的惯例，乍一眼看到这个问题时就会感到困惑。

因此，原始文本的字面翻译不大可能向非专业人士传递太多内容。解决该问题的一种古老方法是译者（或抄录员）添加 34
注释或解说图表：所有重要的数学文本都以这种方式积累了大

量评注；另一种方法是将文本转换为现代的数学符号。想试着解决两位高山旅行者问题的读者，或许会发现这样一来会使解题变得更加明晰。作为一种初步方式，使用现代代数符号可以帮助理解历史上的数学；不过，绝对不应将其误认为是原作者"真正"试图去做的事情，或是他若具有受过良好现代教育的优势就会完成的事情。这样的现代化处理，在最好的情况下仅仅掩盖了原来的方法，而在最坏的情况下则可能导致严重的误解。

比如，现今尚存的公元前2世纪的埃及文本是用僧侣书体写成的。这是一种可以连笔的手写书体，自公元前2000年起便取代了日常使用的象形文字。20世纪初期，这些文本被译成了英文或德文。多年来，这些翻译仍然是标准译本。然而不幸的是，文本内容不仅被译成了现代的语言，而且还被译成了现代的数学语言。比如经常有人说，埃及人过去使用3.16这个值来表示我们如今使用的"π"，也就是通过圆半径的平方给出圆面积的那个乘积因子（用现代的公式可以写成$A = \pi r^2$）。我们在核查这一说法所对应的文本时，却发现埃及人根本不指望读者用任何数字去乘半径的平方。相反，他们指导读者将直径减小九分之一后再平方以求圆面积。用铅笔在纸上简单算一下便可看出，这种方法给出的圆面积是半径平方的$\frac{256}{81}$倍，由此得出了神奇的数值$\frac{256}{81} = 3.16$……不过，"相减再平方"与"平方再乘积"并不一样，即便前者给出了几乎相同的答案，其过程却完全不同，而过程恰恰是历史学家想要理解早期文化的数学思想时需
35 要关注的问题。

翻译巴比伦文本的故事与此类似。这里涉及的语言是苏美尔语和阿卡德语，前者与任何现存的语言都没有关系，后者则是

阿拉伯语与希伯来语的前身；涉及的书体是楔形文字，由削尖的芦苇秆刻在湿润的黏土上写成。20世纪30年代，奥托·诺伊格鲍尔以及弗朗索瓦·蒂罗-丹然翻译并出版了大量数学文本，此后很多年，这项工作一直被认为差不多已经完成了。不过，这些早期的翻译常常将美索不达米亚的计算技术转变为现代代数的等效形式，从而掩盖了原抄录员实际上正在思考以及正在做的事情的真实本质，同时也使计算看起来相当简单。20世纪90年代以来，许多黏土字板才被重新翻译，原始语言也愈加被重视。比如，字面意思是“分成两半”或“附加”这样的词汇传递了物理动作，这些行为信息在诸如“除以2”或“加”这样的抽象翻译中就完全丢失了，而这些词汇使我们可以更好地了解当时理解或教授问题的方式。

阅读并翻译文本只是研究古代数学史的史学家们的一部分工作，尽管是很重要的一项。另一项工作则是在这些文章本身的语境中阐释那些内容。有时这根本不可能：在19世纪发掘或重新发现的许多中东文本，包括几乎所有现存的埃及僧侣书体文本以及数百种古巴比伦楔形文字泥板，都在没有已知出处的古董旧物市场上易手。不幸的是，许多遭抢劫或被偷盗的器物在今天仍然以这种方式进行买卖。

随着我们从古代世界过渡到中世纪，数学文本在脆弱性和稀缺性方面也有了改善，但程度轻微。即便是被慎重保存在图书馆中的文档也不总是安全的。关于冲突时期亚历山大图书馆的毁灭有各种各样的说法，现在还无法证实。当然，它本来就很容易着火，就像任何藏有书籍或手稿的前现代图书馆那样。如 36
今，牛津大学博德利图书馆的读者依然需要宣誓，承诺“不得将任何明火或火源带进图书馆，或在图书馆中点燃，并且不在馆内吸烟”。这是对那段岁月的一个提醒，即这样的举动被证明对人和书籍都是致命的。

我们已经看到约翰·利兰为了记录修道院所属图书馆的图书名录所做的努力，但是当那些图书馆最终遭到破坏而其馆藏流失时，他连馆藏本身的一小部分也无法保存。此外还有其他风险：16世纪，牛津大学墨顿学院在对印刷文本进行现代化处理的同时丢弃了大量手抄本图书。尽管机敏的收藏家们救出了其中一些，但肯定有很多都下落不明。而在1685年，约翰·沃利斯像一个多世纪前的利兰一样，痛苦地抱怨贵重书稿的失窃：他写道，牛津大学基督圣体学院的一部手稿中有两份12世纪的序言“最近（不知被什么人）撕下来，然后带走了”。他希望“不管是谁拿走了序言，都能以某种方式或途径行行好把缺页还回来”，但他的希望却是徒劳：序言仍然不知去向。

私人收藏的纸本同样很脆弱：1644年，约翰·佩尔正担心着不久前刚去世的好友沃尔特·沃纳的数学论文。佩尔写道：

> 沃纳先生的论文中也有不少是我的工作。我很担心他所有的论文都会被收缴，然后以跟数学完全无关的方式被扣押人和债权人瓜分。这些人毫无疑问决心要在他们一生中充当一回占星家，并投票把那些文稿都付之一炬。

同手稿一样，印刷书籍也容易受到火灾、洪水、昆虫甚至人为疏忽的影响。但由于其印刷数量更多，从而更有可能流传于
37 世。不过，传到我们手中的那些不太可能是曾经存世的印本。

某位绅士个人藏书中的一卷昂贵图书要比某个生意人常常翻阅的简明计算手册更有可能保存下来，但至于过去人们实际阅读并用到了哪些内容，这类书籍能告诉我们的却并不多。

构建对历史的真实了解，往往像是在试着拼出一幅拼图，其中大部分碎片都丢失了，而盒面上也没有图片。尽管如此，令人惊奇的是，我们的确拥有存续达数百年甚至数千年的数学著作。其内容在很大程度上纯粹只具历史意义：如今没人能用埃及的分数进行计算，除非作为一种学院派训练；而巴比伦六十进制系统的唯一痕迹，是我们原本就好奇的那种将一小时分为六十分钟，将圆分为三百六十度的划分形式。不过，其他文本经由人们不断地使用和翻译，仍然富有生气，有时甚至可以追溯出一条从过去到现在几乎连续的流传脉络。公认的个例必定是欧几里得的《几何原本》，这份文本已经不止一次被提及，没有它就不可能有完整的数学史。一项有时被称为《几何原本》"传播史"的研究告诉我们许多关于过去的数学思想是如何得以保存、修正并传承的情况。

跨越时间的保存

以上谈到了出自埃及的原始资料的脆弱程度，而同样写在莎草纸上来自古代希腊语世界的文本也是一样的状况。从同时代的引用文本和欧几里得的另一些作品来看，我们估计他的写作年代是在公元前250年左右。但《几何原本》迄今最早的存世文本出自公元888年。这反映出反反复复的抄写持续了一千多年，并且连同所有错误、修正以及"改进"在内。那怎么才能知道我们现今掌握的文本无论在哪个方面都忠实于原始文本呢？

38 答案是，我们根本无从知晓。就《几何原本》来说，我们拥有大量对此书的评论，它们出自此后的希腊作家帕普斯（公元320年）、赛翁（公元380年）和普罗克洛斯（公元450年）等人之手，这些评论告诉我们公元前4世纪或公元前5世纪时文本是如何出现的。这些人比我们更接近欧几里得的时代，但他们距离《几何原本》首次面世的时间也有几个世纪之久。另外，历史学家唯一能接近原始文本的方式，是通过观察比如从一个文本到另一个文本有哪些地方照搬了错误或改动，从而构造存世文稿的“系谱”。他们希望以此方式来回溯出一份“原本”；但这项工作很艰辛，并且对于是否能追溯到一份真实而又唯一的原始材料也没有保证。

《几何原本》最早的存世手稿出自公元888年，是用希腊语写就的，保存于拜占庭。不过，随着伊斯兰教传入地中海通行希腊语的古老地区，该文本也被译成了阿拉伯语。通过与数世纪后罗伯特·雷科德的翻译任务进行比较，人们可以想象早期的伊斯兰译者们可能遇到的困难：作为游牧族群语言的阿拉伯语不大可能包含欧几里得几何抽象概念的现成词汇。尽管如此，阿拉伯语译者却使许多文本免于灭绝。

因此，大多数中世纪的《几何原本》拉丁文译本都是从西班牙或西西里岛的阿拉伯语母本而非希腊语（当时在西欧几乎已经消亡）译成的。我们在上一章中提到的来自巴斯的阿德拉德，便是这样一位译者。12世纪时还有其他几位来自北欧的译者，他们作为学者前往南方游历，访求当地的学问。最终，随着对希腊语的了解逐渐恢复，直接从希腊文母本进行翻译的译本也出
39 现了。

（欧洲）印刷术在15世纪得到应用，欧几里得的《几何原本》终于得以为后世保全下来。它是首批印刷的数学书籍之一，1482年推出的精美版本延续了手稿的制作传统：没有标题页（因为手稿作者习惯在文本的结尾而不是在开头署名），而且还有精美绘制的装饰图案（见图3）。

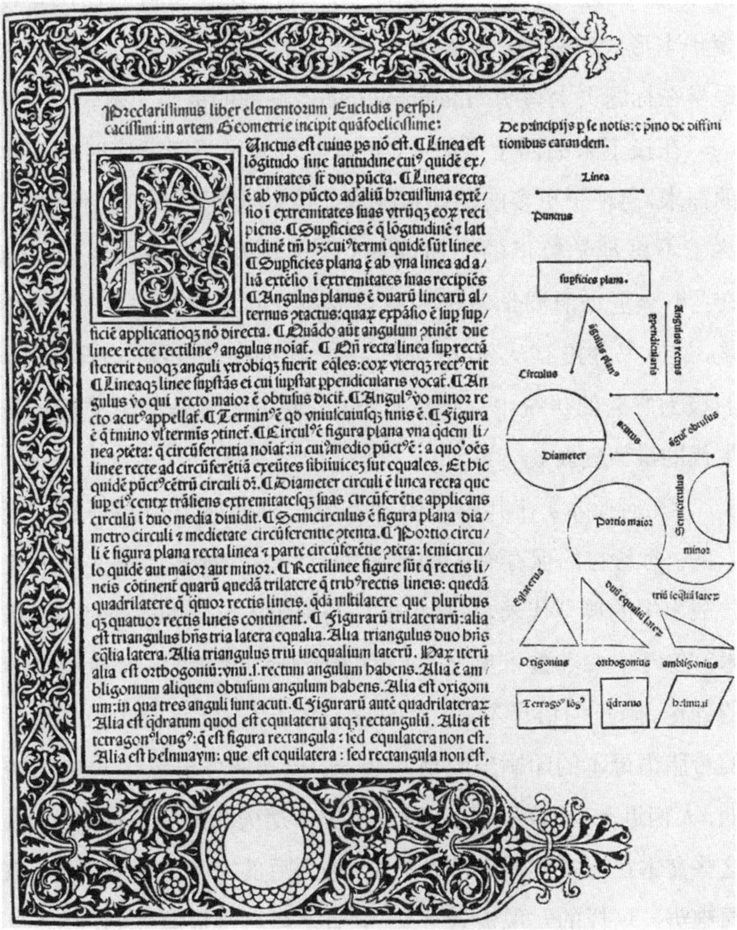

Preclarissimus liber elementorum Euclidis perspicacissimi: in artem Geometrie incipit quãfoelicissime:

Punctus est cuius ps nõ est. ¶ Linea est lõgitudo sine latitudine cui⁹ quidẽ extremitates st duo pũcta. ¶ Linea recta ẽ ab vno pũcto ad aliũ breuissima extẽsio i extremitates suas vtrũqʒ eoꝝ recipiens. ¶ Supficies ẽ q̃ lõgitudinẽ ⁊ latitudinẽ tm̃ hʒ: cui⁹ termi quidẽ sũt linee. ¶ Supficies plana ẽ ab vna linea ad aliã extẽsio i extremitates suas recipiẽs ¶ Angulus planus ẽ duarũ linearũ alternus ꝯtactus: quaꝝ expãsio ẽ sup supficiẽ applicatioqʒ nõ directa. ¶ Quãdo aũt angulum ꝯtinẽt due linee recte rectiline⁹ angulus noiat. ¶ Qñ recta linea sup rectã steterit duoqʒ anguli vtrobiqʒ fuerit eq̃les: eoꝝ vterqʒ rect⁹ erit ¶ Lineaqʒ linee supstãs ei cui supstat ppendicularis vocat. ¶ Angulus vo qui recto maior ẽ obtusus dicit. ¶ Angul⁹ vo minor recto acut⁹ appellat. ¶ Termin⁹ ẽ qd vniuscuiusqʒ finis ẽ. ¶ Figura ẽ q̃ tmino vl termis ꝯtinet. ¶ Circul⁹ ẽ figura plana vna q̃dem linea ꝯtẽta: q̃ circũferentia noiat: in cui⁹ medio pũct⁹ ẽ: a quo oẽs linee recte ad circũferẽtiã exeũtes sibiinuicẽ sũt equales. Et hic quidẽ pũct⁹ cẽtrũ circuli dr̃. ¶ Diameter circuli ẽ linea recta que sup ei⁹ centꝝ trãsiens extremitatesqʒ suas circũferẽtie applicans circulũ i duo media diuidit. ¶ Semicirculus ẽ figura plana diametro circuli ⁊ medietate circũferentie ꝯtenta. ¶ Portio circuli ẽ figura plana recta linea ⁊ parte circũferẽtie ꝯtẽta: semicirculo quidẽ aut maior aut minor. ¶ Rectilinee figure sũt q̃ rectis lineis cõtinent quarũ quedã trilatere q̃ trib⁹ rectis lineis: quedã quadrilatere q̃ q̃tuor rectis lineis. q̃dã mltilatere que pluribus qʒ quatuor rectis lineis continent. ¶ Figurarũ trilaterarũ: alia est triangulus hñs tria latera equalia. Alia triangulus duo hñs eq̃lia latera. Alia triangulus triũ inequalium laterũ. Haꝝ iterũ alia est orthogoniũ: vnũ .s. rectum angulum habens. Alia ẽ ambligonium aliquem obtusum angulum habens. Alia est oxigonium: in qua tres anguli sunt acuti. ¶ Figurarũ autẽ quadrilateraꝝ Alia est q̃dratum quod est equilaterũ atqʒ rectangulũ. Alia est tetragon⁹ long⁹: q̃ est figura rectangula: sed equilatera non est. Alia est helmuaym: que est equilatera: sed rectangula non est.

De principijs p se notis: ⁊ pmo de diffinitionibus earundem.

图3　首印版欧几里得《几何原本》的第1页，1482年

16世纪，《几何原本》的印刷本在短时间内大量出现，首先
40 是拉丁文和希腊文译本，之后是若干种各地语言的版本。罗伯特·雷科德于1551年在《通向知识之路》一书中收录了《几何原本》前四卷中的大部分内容，然后又于1557年在其最后一部出版物《砺智石》里收录了后几卷中其他一些更难的内容。《几何原本》的第一个完整英译本是由亨利·比林斯利于1570年以奢华本形式出版的：该版包含了迪伊的“基础框架”，也是已知最早在标题页上写明“mathematician”一词的英语文本。

在接下来的四个世纪中，随着编辑人员不断适应时代变化的需求，出现了更多的翻译和版本。至20世纪中叶，《几何原本》终于不再是学校里的一门课程（尽管其内容依旧在课程中存在：小学生仍旧要学习构造三角形以及对角进行二等分）。但是，这本书并没有从公共领域消失。在翻译与改编《几何原本》的极为悠久的传统中，现代交互式的网页版本是适用于每一新生代的最新创新。

《几何原本》在其影响范围及流传时间两方面都是独一无二的；而与它的保存有关的故事则是许多其他希腊语文本的流传故事中的典型代表，包括启发费马提出大定理的丢番图的《算术》在内。大多数古典文本都有着类似的故事，内容涉及早期的评论、阿拉伯语译本、之后的拉丁文译本，以及最终源自幸存的希腊语母本的印刷出版物。唯一的一次例外出现在20世纪初期，人们近乎奇迹般地重新发现了原本遗失的阿基米德的文本。这些文本在一本拜占庭祷告书中，被之后的文字与绘画覆盖，隐约难辨。这样的发现极为罕见，并且足以再次提醒我们，有多少来自不同文明的数学成果早已遗失。

跨越空间的保存

尽管书面文本比较脆弱，数学的流传却不仅跨越了时间，有时也跨越了空间，有时则兼而有之。我们不妨从一个谜题开始。41
大英博物馆保存了一块古巴比伦的黏土板（编号BM 13901），上面有一个问题的开头是这样的：

我把正方形的面积与边长相加，使和为0; 45。

用本章前面提醒过的那种技术，让我们恰当地引入代数符号来看看问题所在。假如我们令正方形的边长为s，那么其面积便是s^2。数字0; 45是一种现代转写，我们可以将其解释为45/60或3/4。因此，上面那个陈述用现代术语就可以写成等式$s^2 + s = 3/4$。巴比伦用于求正方形边长的技术涉及对几何形状进行切割并重新排列；对于训练有素的从业人员，可以将其归纳为一系列简短的说明，一种可以确保给出答案的秘诀。

现在，让我们考察下面这个问题。它出自一篇主题为“还原与平衡”的文本，由巴格达的花剌子米在公元825年前后撰写。

1个正方形加21个单位长度等于10个根。

这里的“根”是指给定正方形的平方根。因此，我们如果再一次使用现代符号，那么就会发现问题可以写成$s^2 + 21 = 10s$。换句话说，这与两千五百多年前被记录下来的那道古巴比伦问题密切相关。此外，花剌子米给出了一个非常相似的方法来求解。他的著作影响极大，以至于今天“代数”这门学科的名称就来源于他。

同样的问题和同样的解答，数百年后再次出现在世界的同一地区。这是巧合吗？没有任何证据可以表明其在这么长时间跨度上的连续性，就像我们对欧几里得《几何原本》所掌握的那样；当然，在古代伊拉克或伊斯兰化的伊拉克境内也没有相关证
42 据。不过，我们确实有证据表明思想从晚期的巴比伦文明传播到印度，而数学后来又从印度传回到巴格达。像此处讨论的那些问题很可能是传播中的一部分：我们不能认定而只能推测。但不管怎么样，此处都值得详述一下我们所了解的事情中更有把握的那部分。

自大约公元前500年至公元前330年，古代伊拉克同印度西北部还是波斯帝国境内的遥远伙伴。此后，该区域一度被纳入亚历山大大帝治下。巴比伦的数学被印度吸收，这方面有详尽而相当清晰的证据，尤其是在天文计算中：这可以从印度使用基数60来度量时间与角度，并用类似的方法来计算全年的日照时长看出。（与其他早期社会一样，在印度，用于仪式及其他目的的正确计时至关重要。）后来，希腊语的天文或占星术文本有了梵文译本。因此，用于测量天文高度的希腊词汇“chord”（弦）成了印度“sine”（正弦）的基础。早期印度文献的匮乏，使我们无法了解另有哪些知识确凿地传向了东方或其他方向：例如，有一些记有天文方面内容的黏土碎片，来自伊斯兰化之前的伊朗，从而表明了梵文文本在那里的影响。

至公元6世纪末（甚至再早些），印度中部地区就已经发展出一种仅有十个数字且与位置值体系连用的记数系统。其重要

性怎么说都不夸张。用现代的话来说，这意味着我们仅用“0、1、2、3、4、5、6、7、8、9”这十个符号就可以记录任意大小的数字。所谓“位置值”，意味着比如“2”和“3”在200003和302这两个数字中代表不同的值，因为它们所处的位置不同。在上述两个数字中，零都用作占位符，这样我们就不会将200003误认作23或者把302误认作32了。一旦理解了这一点，就可以将几个相同 43
的加法与乘法规则应用于任意大小的数字。当然，历史上还有许多其他的记数方式。但随着数字变大，所有这些方式都需要发明越来越多的新符号，而且没有一种能便于用笔在纸上进行计算：让我们试着对两个用罗马记数符号表示的数字求和，比如xxxiv和xix，并且不将其转写成我们更熟悉的形式。

早在公元7世纪，在柬埔寨、印度尼西亚及叙利亚的某些地区，人们就已经知道了后来被称为印度记数法的那套符号系统：对此，比如叙利亚的主教塞维鲁·塞博赫特就曾对此给予高度评价。至公元750年，伊斯兰教已经遍布（并超越了）古波斯帝国的领域范围；至公元773年，印度记数符号随着敬献给哈里发曼苏尔的天文专著从印度来到巴格达。大约在公元825年，花剌子米，就是我们前面提到的那位代数领域的作者，写了一篇关于如何使用印度记数符号的文章。原始文本已经失传，但其内容可以通过后来的拉丁语译文复原。文章首先讲如何用阿拉伯语形式而不是梵语形式来写十个数字，并仔细解释位置值以及零的正确使用；接下来是对加减法、翻倍与减半、乘除法的说明，一些关于分数的教学（包括六十进制分数在内），以及提取平方根的说明。花剌子米的著作为算术文本的形式创立了长达数百年的范式：尽管在当时文献资料已经大大扩展，在许多17世纪的欧洲文本中仍然可以轻松辨别出其框架。但就眼下来说，让我们还是回到印度记数符号本身，或者来关注随着其继续向西传播进而演变成的印度–阿拉伯数字。

至10世纪末，这种记数符号从印度传到了地处伊斯兰世界另一端的西班牙，并习得了预示着现代西方记数系统的西阿拉伯形式，而不是如今在通行阿拉伯语的国家中仍然使用的东阿
44 拉伯形式。这些记数符号由西班牙逐渐向北传播到法兰西和英格兰。关于记数符号的一种传说是，它们由一位叫热尔贝的僧侣传入基督教欧洲。此人也就是后来的教皇西尔维斯特二世，他在公元970年前曾到访过西班牙。热尔贝确实在算盘计数器上用过记数符号，但仅凭这样一条微不足道的证据，人们几乎不会把将符号系统引入欧洲其余地区这件事归功于他：我们不清楚他是否学会了相关的计算方法，还是仅仅把数字充作装饰符号；我们也不知道他的算盘在多大范围内为人所知或得到使用；此外，肯定还有其他前往西班牙的旅行者，他们同样带回了一些关于记数符号的知识来向朋友们展示。有关记数符号的知识可能只会以一种零散的方式缓慢地传播，直到人们开始更好地认可其实用性。

我们很清楚，出自西班牙的天文表（即“托莱多天文表”）于1140年及1150年分别为马赛和伦敦改进过。天文表的使用说明是从阿拉伯文译成拉丁文的，但其本身却并没有翻译：有谁会想要把测量度、分、秒所得的两位数的数列转换为麻烦的罗马数字呢？正如天文表将印度记数符号带到巴格达一样，这套符号后来又被带到了北欧：对天文学家来说，数字不仅有用，而且对理解他人的观测至关重要。

在更世俗的层面上，数字的知识和相关的计算方法也必须通过贸易向西和向北传播。比如，十字军自11世纪末起便会接触到它们。然而，与天文表不同，买卖记录是短暂的，并且早已不见踪影。

到了12世纪，有人专门编写了一些文本来解释新的记数符号及相关的计算方法。其中之一便是斐波那契的《计算之书》， 45
这本书流传于意大利，却并没有在北欧传播。在法兰西和英格兰，反倒发现了被称为“algorisms”的拉丁语文本。这个名称转写自文本开篇的词汇“Dixit Algorismi”，意思是“花剌子米如是说”。就像花剌子米的原著那样，这些文本介绍了如何记写数字以及如何将其用来进行基本的算术运算。来自法兰西北部的亚历山大·德·维尔·迪厄用诗歌形式撰写了一篇格外迷人的作品，被称为《算法的卡门》。开头几行的译文如下：

> 当下的这种艺术被称为“algorismus”，
> 我们在其中用上了十个印度数字：
> 0、9、8、7、6、5、4、3、2、1。

亚历山大接着解释每个数字的位置如何重要：

> 如果您将其中任何一个放在首位，
> 它仅仅表示自己；如果在第二位，
> 则表示自身的十倍。

尽管数字有着明显的优势，但其真正被投入使用的进程却很缓慢。这并不是因为有时人们所以为的，它们来自东方，有着非基督教背景的起源，而是因为对于日常应用，旧式的罗马记数系统辅以手指或记数板来进行计算就足够了。另外，并非所有人都觉得新的记数符号易于学习：直到14世纪或15世纪，意大利卡文索本笃会修道院的一位僧侣从第30章起才将章节编号记为XXX、XXX1、302、303、304……不过，印度-阿拉伯数字最终取代了其他所有的记数系统，一旦它们向西被带到了美洲，就几
46 乎完成了其环球之行。

关于数学得以远距离传播的方式，还有其他的故事。比如，中国的所有邻国都学习了中国的传统数学。直至17世纪耶稣会士带着欧几里得的《几何原本》进入中国，中国传统数学才有了与西方的交流。而且毫无疑问，中国传统数学与印度也有所交流。这种活动在近代一直持续着：19世纪中，欧洲数学从法兰西和德意志这两个中心地带向外传到了欧洲的外围，一端是巴尔干半岛，另一端是不列颠；然后又传至美国，最终传到了世界各地。这种成规模的传播是现代特有的，但数学中的思想却已经流传了很长时间。

不要忘记人在其中的作用

本章中，对于过去的某些数学是如何历经漫长的岁月甚至有时跨越遥远的距离幸存下来的，我已经进行了叙述，虽说形式比较零散。不过，我已经试着谨慎用词。用来表达数学思想传承的一个常用词汇是“transmission”（常义为“传递、传播”），但我不喜欢这个词：它除了隐含“无线电天线”这层意思外，还暗示着数学思想的创始人故意将他们的想法及发现指向了后人。

这种情况实际上极少发生。把数学写下来，多半是为了个人自用或是写给自己同时代的人。能远远超越这种目的而幸存下来，在很大程度上取决于环境。同样，我也尽量避免谈论思想只是在单纯地传播、扩散，就好像它们是自身拥有力量的庄园杂草一样。 47

恰恰相反，每次数学交流，无论大小，都是由人为因素引起的。在上述长期流传的故事背后，是无数微小的相互影响与交易。我们已经瞥见了其中的一些：印度的使节们向巴格达的哈里发做着自我展示；拜占庭的誊写员抄录着自己可能几乎看不懂的手稿；佛罗伦萨的商人们在亚历山大的市场里讨价还价；还是在亚历山大，一千年前的某位图书馆管理员仔细列出了他所负责的书卷，他或许像后来的约翰·利兰一样，也对图书的损毁感到不安；费马把信寄给了牛津的沃利斯，却如石沉大海，杳无回音；怀尔斯在一次演讲中首次公开了他的证明，后来又通过电子邮件发布了关于证明最终修正的消息。数学思想之所以四处传播，只是因为人们在思考，在与他人讨论，在记录并保存相关的文档。没有人，数学思想的传播就根本无从谈起。 48

第四章

学习数学

一个容易忽视的事实是，现代社会中学习数学的最大群体不是成年人，而是小学生。世界上任何一个地方的年轻人，若足够幸运而能接受教育的话，就有可能会花费大量的时间来学习数学。在发达国家，每个学校每周会有两到三个小时用于数学，并可能持续十年或者更长的时间。

有鉴于此，回想起下面这一点，即把数学纳入学校课程是现代的一个现象，就有些令人惊讶。比如1630年前后，约翰·沃利斯既不是在学校也不是在剑桥学习算术，而是师从他在学习做生意的弟弟。他后来成了牛津大学萨维尔几何学教授。三十年后，塞缪尔·佩皮斯在努力学习乘法表。他也曾在剑桥接受教育，还是海军委员会的成员，以机敏过人且有素养著称。尽管如此，在大多数文明社会中，将数学知识至少传给下一代中的一些人，已经被视作一项重要任务。

数学在哪些方面被认为是有用的，以及用于哪些目的，一项
49 对所教授内容及方法的研究告诉我们很多这方面的信息。本章

我们就要来检视两个案例研究，因为我们掌握了相对较好的材料：一个发生在公元前1740年之前某段时间在伊拉克南部尼普尔的一间教室，另一个发生在公元1800年之后不久位于英格兰北部坎布里亚的格林罗学院里的一间教室。

巴比伦的一间教室

尼普尔古城位于幼发拉底河的沼泽地带，大约在现今巴格达与巴士拉两座城市中间的位置，曾经是重要的宗教中心，围绕着一处献给恩利尔神的寺庙群而建。像后来中世纪欧洲的大修道院和僧院一样，巴比伦的神庙得到了大量供品，并且掌控了土地和劳动力，因此需要训练有素而能处理书面记录与计算的抄写员。有些孩子注定要从事那些通常由家庭来经营的行业，他们可能很早就开始接受培训。

尼普尔有一处不大的砖砌房子，如今被称为“F号房屋”，看起来曾经是这座城市中可能有的几所抄写培训学校中的一所。该处房屋始建于公元前1900年之后的某个时期，靠近献给伊娜娜女神的神庙，并在公元前1740年这一年的前不久被用作学校。像所有砖砌结构一样，它需要定期维修。这里不再作为学校之后，又进行了第四次或第五次重建。这一时期，工匠们充分利用了数百块废弃的教学用黏土板，把它们整合到了新房子的地板、墙壁和家具之中。其他一些部分损坏的黏土板也已经被发现，它们同回收箱中大量未曾使用的黏土混在了一起。

这处房子被用作学校时，分成了三四个内部空间和两个庭院，庭院里有些长凳和回收箱。不幸的是，我们不知道那些学生 50

的姓名或年龄，也许一次最多只有一两位；我们也不清楚他们占用庭院里长凳的次数或时长。不过引人注目的是，他们使用黏土板的方法已经能使楔形文字研究者重构当时的课程。

出自这处房屋的许多黏土板，一面（正面）是平的，另一面（背面）则微微凸起。正面左侧是老师写的范文，右侧是学生的摹写。在黏土板微微凸起的背面有更长的段落，是之前所学的课件材料。重新书写要么是为了进一步练习，要么也许是用作记忆测试。这样的尼普尔黏土板大约有一千五百块，每件上都有“较早”和“较晚”的材料。尼克·维尔德维斯在20世纪90年代就能够从中辨识出学校基础课程中的一种连贯顺序，比如从基本的写作技巧开始，到苏美尔文学的开端为止。埃莉诺·罗布森对出自该处房屋的大约两百五十块类似的黏土板运用了相同的方法，以便能对学校课程进行同样的研究，从而发现数学在其中的地位。

学生们需要迈出的第一步是，学习楔形文字符号正确的书写技法并组合起来拼出自己的名字。之后他们借助单词表来学习书面词汇，比如从树木和木制品开始，接下来是芦苇、器皿、皮革及金属制品，动物和肉类，石头、植物、鱼、鸟及服装，等等。在这里通过测算船只的容量、树木和石头的重量、芦苇制测量杆的长度，就已经介绍了一些数学词汇。随后，更多的度量衡单位也出现在专用的重量和度量列表之中。

接下来，学生被要求熟记逆算表（成对的数字相乘得60）
51 以及二十多个标准乘法表。例如某个逆算表可能以这种形式展开：

2	30
3	20
4	15
5	12
6	10
8	7 30
9	6 40
10	6
12	5

（在我们如今仍然用在小时和分秒计算上的六十进制算术中，7 30等于$7\frac{1}{2}$，而6 40是$6\frac{2}{3}$。）乘法运算表需要相当多的记忆。比如用于16 40的乘法表是这样展开的：

1	16 40
2	33 20
3	50
4	1 06 40
5	1 23 20

据估计，在完成学校其他练习的同时来学习全套运算表可能需要一年的时间。在这一阶段，学生也开始用苏美尔语书写完整的句子，其中一些语句包含了之前学过的度量衡单位。

只有在完成以上所有之后，也就是学生们习得更高级的苏美尔语之后，他们才开始脱离标准运算表来进行自己的倒数运算。在这处房屋为数不多的“高级”黏土板中，有一块上面记录了一些用于求解17 46 40倒数的运算方法（答案是：3 22 30）。
52 这些内容与一段文学作品选摘写在了同一块黏土板上。这部文学作品被称为“导学对年轻抄录员的建议”，其中包括以导学自己当学生时的记忆为基础的一些道德说教：

> 就像弹起的芦苇，我一跃而起，投入工作。
> 我没有偏离老师的指导；
> 我没有主动开始进行练习。
> 指导老师对我的作业很满意。

出自该处房屋的大多数高级文本都与数学无关，而是像“导学的建议”那样的文学作品。不过，其中有许多都提到在社会公正管理中读写能力与运算能力的应用。在一首献给抄录员的守护神尼沙巴的赞美诗中有这样几句，赞美她将礼物赐予国王：

> 一根单杆的芦苇和一条用于青金石的测量绳，
> 一把标尺和一块带来才智学识的书写板。

位于坎布里亚的一间教室

1780年，约翰·德雷普在英格兰西北海岸的希洛斯创建了格林罗学院，就在与苏格兰交界处以南几英里的地方。像尼普尔那所在“F号房屋”中的学校一样，格林罗学院也是某种家族事业。他的父亲约翰·德雷珀此前曾在同一海岸再往南三十英里的怀特黑文办了一所学校。这所位于怀特黑文的学校当时注重与“贸易及航海技术”相关的学科，德雷珀则出版了两本教科书供他的学生们使用：一本是1772年出版的《青年学生的口袋书》，包括算术、几何、三角学及测量学，为提高在校青年的水平而编排；另一本是1773年出版的《领航员手册，或一套领航技术的完备体系》。德雷珀于1776年过世；约翰继承了他父亲的图书、数学工具及一些财产，这样他便有能力在几年后来创建格林罗学院。1795年德雷普去世。此后学校由另一位家族成员约瑟夫·索尔打理，他是德雷珀妻子的亲戚，一直管理该校近五十年。课程范围也有所扩大，新增了包括希腊语、西班牙语以及 53
《圣经》研习在内的课程。但格林罗学院与位于怀特黑文的那所学校一样，仍旧将数学教学视为重点。

这所学校不仅吸引了当地的男孩，还吸引了英格兰其他地区甚至海外的男孩。九岁大的孩子就可以注册入学，甚至一度六岁大的孩子也可以，但有时也有二十来岁的年轻人在那里接受教育。不过，大多数学生都是十四岁或十五岁。1809年的记录显示，十一岁大的罗兰·考珀是年龄最小的学生之一，二十三岁的詹姆斯·欧文是年龄最大的学生中的一位，他们都要完成相同的基础课程，包括英语、写作和算术。其他大多数的男孩还要学习绘画、法语或拉丁语，以及非常宽泛的数学专题。十五岁的约翰·科尔曼所学的课程就非常典型：英语、法语、写作、绘画、算术、几何、三角学、测定、测量、簿记、球面学、天文学、力学、代数以及欧几里得几何。学校开设的其他数学专题还有日晷制作与计时（日晷的结构）、测量估算及防御工事。而十六岁的乔治·皮特看起来非常出色，还参加了圆锥曲线和“流数”（牛顿版本的微积分）的课程。

不过，我们很幸运，能从格林罗获得更多的信息，而不仅仅是科目的列表。数学教育家约翰·赫尔西于2005年去世，在此之前，他收集了两百多份数学抄本，它们出自1704年至1907年间英格兰及威尔士各校学生之手。这些不是现代意义上的练习本，学生们并没有浪费宝贵的纸张来一遍遍地练习相似的问题。相反，他们仔细地记下标准问题的翻版示例，从而为自己确立了一系列可行的题例，这样就能便于日后应用。许多题例取自当时流行的教科书，尤其是弗朗西斯·沃金盖姆那本广为流传的
54 《教辅助手》（首版于1751年）。不过，其他许多例子肯定是老师自己特别为学生们设计的。

赫尔西的搜集包括罗伯特·史密斯于1832年至1833年之间撰写的五本数学工作手册（见图4）。在这两年中，罗伯特记下了近一千七百页的数学题例，所以我们就对他所研究的内容有了非常详尽的了解。这些并不是罗伯特所作的第一本书，因为它已经远远超越了基本的加、减、乘、除运算。现存最早的一本来自1832年，以“三的法则”开篇。这条法则让不知多少代的学生能回答诸如下面这样的问题：数量为A的一组人在B天内挖了一道沟，那么数量为C的一组人做同样的工作需要多长时间？这条法则之所以如此命名，是因为必须从三个已知量（A、B、C）中求解第四个量（即答案）。这条法则起源于印度，可能是随着印度数字系统一并西传：其在伊斯兰及欧洲的算术文本中普遍存在达数世纪之久。

学生们曾以死记硬背的方式学习过“三的法则”：也不指望19世纪中的某个英国男学生会“主动开始进行练习”，就像他们的巴比伦前辈那样。在上面的示例中，他会学到需要用B乘以A

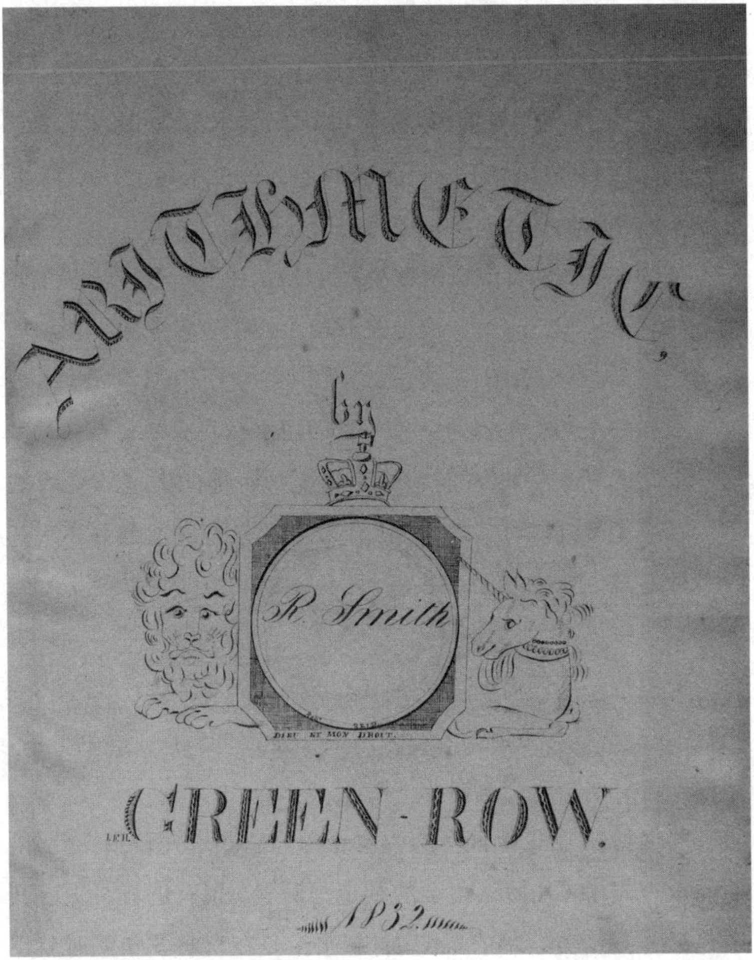

图4　罗伯特·史密斯所著算术书的首页，格林罗学院，1832年

再除以C来求得正确的答案。不过，当然总有些变化来难倒粗心的学生：罗伯特·史密斯就必须学习“三的正向法则”、“三的逆向法则”以及“三的双重法则”。紧接着这些专题之后，便是易货贸易、利息、团契规则（利润分配）、普通分数、十进位分 55

数，以及算术级数与几何级数，此外还有另一些内容。他的第二本书显然是在同一年编写的，研究工作贯穿一系列类似的主题，同样以“三的法则”开篇，以级数和十二进制结尾。这些书看起来是按先后顺序写就的，因为罗伯特本人标上了卷一、卷二。目前还不清楚为何他要对类似的题材研究两次。

56 他的许多题例都取自沃金盖姆。比如，有两个关于置换的示例，下面是其中之一（另一例是关于12个铃铛有多少种按法）：

> 一个年轻人因为一家不错的图书馆提供的便利来到镇上。他与房东讲价，付给对方40英镑的膳宿费，只要他每天在晚餐时可以将家人（除他自己外还有6人）安排在不同的位置，就可以一直住下去。问：他花了40英镑可以住多久？

罗伯特在问题后面直接给出了正确的答案（即$1 \times 2 \times 3 \times 4 \times 5 \times 6 \times 7 = 5\,040$天）。不过，之后他立即转向了普通分数，在这一点上他与沃金盖姆非常接近。

罗伯特在1832年撰写的两本算术书总共近九百页。此外，他的第三本书《几何三角学的测定与测量》也有近五百页。其中包括一些精美绘制的草图，看起来像是在格林罗学院时就已经受到了启发（见图5）。

下一本书的标题页是“算术，罗伯特·史密斯撰，1833年于格林–罗”，主题是“关于普通规则的实践问题”。被称为“包裹单”的那些问题让人特别感兴趣，因为学生们经常用自己的名字
57 和日期来代替沃金盖姆原有的。因此，罗伯特的第一张包裹单是这样的：

图5 罗伯特·史密斯提出并解答的一个三角学中的问题，格林罗学院，1832年

格林-罗，1832年7月13日

托斯·纳什先生

买了罗伯特·S.史密斯

8双精纺毛料长袜，每双4先令6便士，计1英镑16先令

5双线料同上，每双3先令2便士，计15先令10便士

其他包裹单上更多的日期贯穿1832年的整个7月，并持续到了8月，这表明罗伯特在1832年可能就已经着手编写了，也就是说他并非于1833年才开始写这本书，而恰恰是完成于这一年，即书名页上的日期。托斯·纳什这个名字在罗伯特的第一本书里除了别的地方还出现在结尾处，排在一起的还有另一个人名：罗伯特·里德。这暗示两人可能是罗伯特的老师。这位里德又一次出现是在如下这类包裹单中：

18码优质蕾丝花边，每码0英镑12先令3便士，计11英镑0先令6便士

5副优质儿童手套，每副2先令3便士，计11先令3便士等等。

58 罗伯特在1833年完成的第二本书关于“几何体的测定”。这本书里有五种常规几何体（即四面体、六面体、八面体、十二面体和二十面体）体积与表面积的复杂计算，还有砌砖工、石匠、木匠、铺石板匠、油漆工、玻璃匠、水管工以及其他行业所用的典型计算，并且用到的都是适用于各行业的单位。比如，罗伯特了解到，油漆工是以平方码为单位来估算“踢脚板、门、百叶窗”的面积，但“一直都得去掉壁炉及其他开口的面积”。

遗憾的是，我们不清楚罗伯特着手以上这些事情时的年龄；但我们可以看到，在格林罗的那些岁月带给他一种既有理论性又完全可实践的数学教育。

女孩们

我有些犹豫地加入了这一节内容。它将构成人类半数的一群人处理得就像少数群体一样；但并没有偏离这样一个事实，即就大多数社会的大部分历史来说，人们认为没必要甚至不宜对女孩进行教育，当然也就谈不上对女孩在数学或科学等方面的教育了。所以毫不意外，很少有著名的女数学家，就像在最近几个世纪以前很少有女性作家、律师或医生一样。这种状况一定让成千上万甚至更多的聪明女性有些沮丧。尽管如此，时不时还是有人得到了或者为她们自己创造了接受数学教育的机会。

这些女性中有这样一群人，她们拥有大量的财富或者有足够的闲暇来沉迷于所选择的研究。早期的一个例子是公元1世纪末中国的邓皇后，她曾经上过算术课。这一时期不同寻常的 59
是，她的老师也是一位女性，叫班昭。很久之后，到了17世纪40年代，波希米亚公主伊丽莎白和瑞典女王克里斯蒂娜都上过笛卡尔的课，尽管相比于数学，她们可能对他的哲学思想更感兴趣。一个世纪后，欧洲最多产的数学家莱昂哈德·欧拉写了两百多封关于数学与科学主题的书信给安哈尔特-德绍公主。这位公主是普鲁士腓特烈大帝的侄女。这些信函有法文版、俄文版和德文版，后来又以《写给一位德意志公主的书信》为标题出了英文版；时至今日，仍在出版发行。

然而，普通女性进入数学领域的一条更为普遍的途径是师从父亲、丈夫或兄弟。例如，公元前19世纪，在巴比伦锡普尔镇有两位女抄录员，伊纳娜-阿玛加和尼-娜娜姐妹，看起来她们很有可能是从父亲阿巴-塔布姆那里学到了专长，他也是位抄录员。两千年后，邓皇后和她的兄长们从父亲那里接受了最初的教育，尽管他们的母亲似乎认为这对一个女孩来说是浪费时间。邓皇后后来的老师班昭是学者班固的妹妹，她对兄长的工作非常了解，在兄长去世后续补了他未完成的遗作，包括一篇关于星象的论述。数学领域最著名的父女也许是公元4世纪末亚历山大的赛翁和希帕蒂娅。不过，我们手上没有出自希帕蒂娅本人的著作，只有一些关于其生活与离世的二手记载。围绕着其生死汇积了许多传奇。

女孩在家庭中接受教育的形式一直持续到近代的早期。17世纪70年代，约翰·奥布里在文章中谈到了以前的朋友爱德华·戴夫南特，后者是多塞特郡吉林厄姆的一位牧师。奥布里记述了这位朋友对数学的热爱，尽管“他作为牧师不乐意出书，

60 因为世界不应该知道他是如何度过大部分时光的”。戴夫南特不仅教奥布里代数，还教授自己的女儿：

> 他非常乐意教学并进行指导。他帮助了我，首先给我讲授了代数课程。他的女儿都是代数学家。

碰巧的是，我们知道爱德华·戴夫南特用代数的方式向自己的长女安妮教授了些什么，因为热衷于记录人类所有事务的那位奥布里在1659年抄下了她笔记本里的内容。安妮生于1632年（这是她妹妹凯瑟琳的出生年份）之前，并在1650年同安东尼·埃特里克结婚，因此她很可能在17世纪40年代初期被训练成为“代数学家”。奥布里对安妮作品所做的抄本是这样开头的：

这些代数上的内容是我从安妮·埃特里克那里抄录的，她是戴夫南特的长女，戴夫南特博士是一位非常出色的逻辑学家。

安妮笔记本中日期在前的那些问题，以及她书写时所用的拉丁文，对年轻的初学者来说是典型的。比如其中有这样一题：一位年轻男士出现时，一群女孩正在散步。男士（用拉丁语）说道："12位姑娘，你们好啊。"其中一个女孩（也用拉丁语）立即回应道："如果用我们的人数乘以5，那么我们人数上多出12的数量就同我们现在少于12的数量一样了。"读者不禁要问：那里有几个女孩？几页之后，我们发现安妮正在做一个题例，其形式与解答是巴格达的花剌子米在八个世纪以前提出的：什么数乘以6再加16会得出自身的平方（用现代符号即$6x + 16 = x^2$）。终于，在笔记本结尾处，拉丁文与数学的水平都变得更加成熟了。其中倒数第二个问题直接出自丢番图的《算术》：把370分成两个正立方体，其根都是整数且相加之和为10。就像安妮所能展示的那样，答案是7^3加3^3。这个题例中的数字都经过精心筛选，好给出一个简单的解答。但要用一个完美的立方体来替换370这个数字却是不可能的，正如远在图卢兹的费马几乎同时所发现的那样。 61

时间一直来到18世纪，女孩们却只有坐拥社会地位上的优势或有宽容通达的父母，才有可能被教授数学，就像邓皇后和安妮·戴夫南特那样。在费马大定理上取得进展的那些关键人物中，有一位叫索菲·热尔曼，她同时受益于上述两个方面。1776年，她出生在巴黎一个富有且受过教育的家庭，法国大革命爆发时她才十三岁。由于被限制在家中，她只好在父亲的图书馆里自娱自乐，就这样发现了数学这一科目。她的父母起初认为这一学科并不适合她，不过后来在她的决心面前，他们的态度软化了。十八岁那年，她设法从新成立的巴黎综合工科学校拿到了讲义。尽管她本人未被允许进入课堂，但她化名“勒布朗先生”把作业交给了约瑟夫-路易·拉格朗日。他是该校最伟大的老师之一。四年后，她再次用勒布朗这个化名与同样伟大的德国数学家卡尔·弗里德里希·高斯通信。即使在察觉了她的真实身份之后，拉格朗日和高斯仍然赞赏她的数学能力并钦佩她的勇气。她是值得二位称赞的。索菲一生中大部分时间都在与不平等作斗争：有着同样天赋的男孩可能享有的那种教育，她从未接受过；而她的工作成就却常常被错误和不完整所破坏。她从来没有担任过任何正式的职务。尽管如此，1831年她去世之后，高斯指出，她本该在哥廷根获得荣誉学位。哥廷根是当时欧洲主要的数学中心之一。

文章或海报提到“数学中的女性”时总是将希帕蒂娅和索菲·热尔曼作为典型；不幸的是，这不是因为她们代表了各自的时代和地域，恰恰相反，她们并非所处时代与地域的代表。像班昭或安妮·戴夫南特这样被埋没的女士，大体上更能体现女性在数学及数学教育方面的经历。

至19世纪，随着西欧的女孩们开始更多地受益于基础教
62 育，她们的境况慢慢改善了。赫尔西的藏书中，女学生的抄写本很少。不过，我们所掌握的那些，让我们对英格兰和威尔士的各所学校面向女学生的数学教学内容有了一些有价值的了解。

1831年，也就是罗伯特·史密斯在格林罗开始撰写上述几

本书的前一年，埃莉诺·亚历山大在费尔沃特学校里正研究着“还原”（“把30英镑1先令$1\frac{1}{4}$便士换算成法寻”）及“三的法则”（“假如7码布料值3英镑10先令，那么65码要多少钱？”）。这所学校位于南威尔士纽波特以北的一个山谷中。她的整个抄本有一百二十七页，就由这两类问题组成。三年后，即1834年10月初，安·威特曼在约克郡附近的阿普尔顿-勒-穆尔斯开始通过自己的方式来研究沃金盖姆的《教辅助手》（见图6）。她早期的学习记录都标有日期，所以我们知道她在“简单加法”上花了大约十天时间，而对“乘法”却用了整整一个月。圣诞节后，她学习了（用于货币、布匹尺寸、土地丈量、啤酒和麦芽酒度量等方面的）“复合加法”；3月底，她就到了“包裹单”这一主题。在她的题例中，8双精纺毛料长袜到了威廉·G.阿特金森夫人（她的老师？）手上，而亨利·威特曼先生（她的父亲？或她的兄弟？）则买了15码缎子。1837年4月她开始了“练习”，这种方法依赖于掌握标准重量和度量的分数（见图7）。翻看完二百五十页之后，威特曼的抄本结束于1837年5月10日这一天，此时她已经以自己的方式学到了沃金盖姆的“复合利息”。她在不到三年的时间里记下的数学内容，比罗伯特·史密斯几个月内完成的篇幅要少得多，不过数量仍然相当可观。

二十年后，林肯郡斯坦菲尔德的伊丽莎白·阿特索尔也通读了沃金盖姆的书，从“复合加法”到“三的法则”（比如“3磅咖啡要1英镑1先令8便士，那么29磅4盎司咖啡得花多少钱。”）。在她的题例中，1850年10月22日有8双精纺毛料长袜到了查普尔夫人手里。不过，1861年时I. 诺曼小姐在曼彻斯特 63 的休姆以英格森先生冠名的多塞特街学院里的活动却略有不

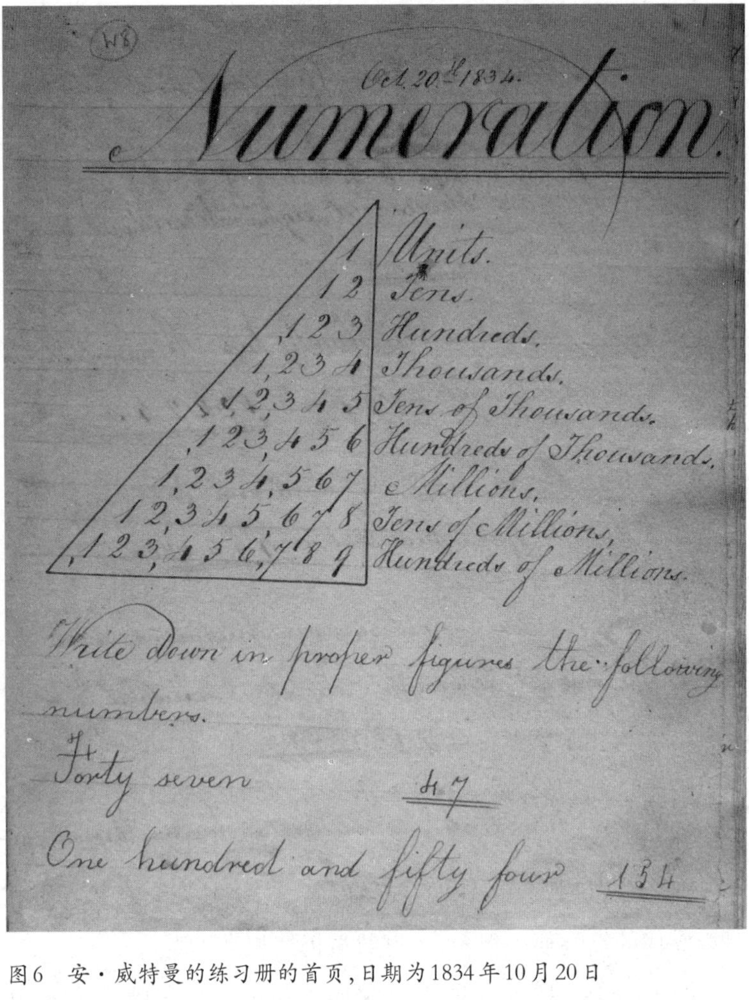

图6 安·威特曼的练习册的首页，日期为1834年10月20日

同。她的抄写本专门印了学校的名字，就在淡蓝色的纸上，页边还画有双红直线。诺曼小姐在第1页记下了“累进算术”，但可惜没有取得多大进展：这个抄本有六十页，每一页都写满了对英镑、先令和便士的乘法或除法运算（比如“一码布料要$7\frac{8}{9}$便士，那4767码我得付多少钱？”）。一年后，诺森伯兰郡卡希尔德学

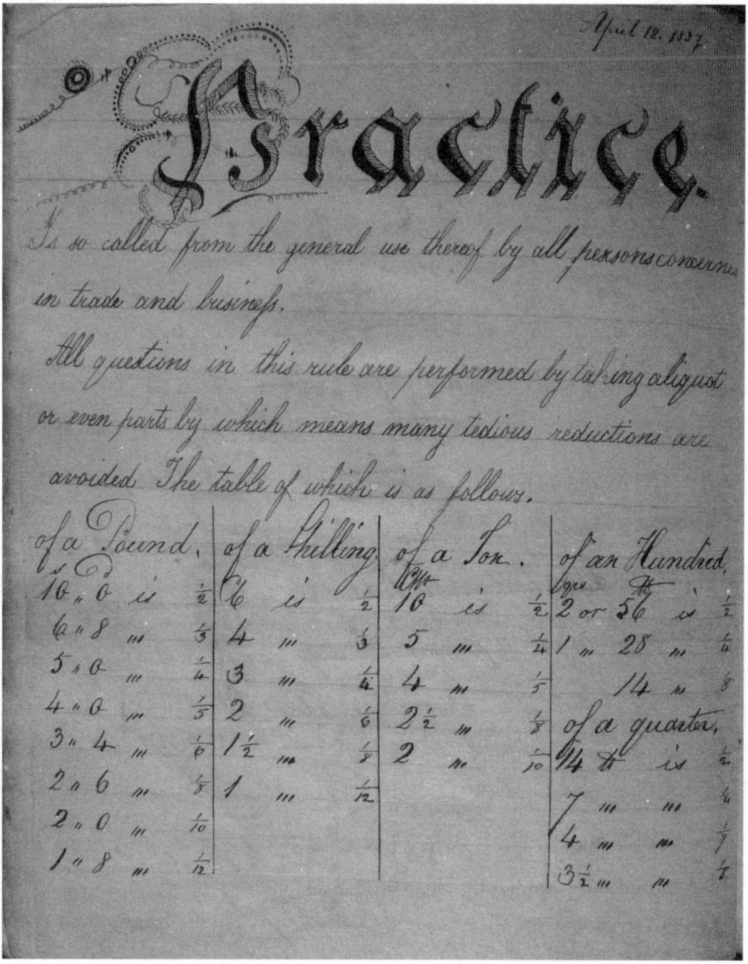
April 12. 1837

Practice

Is so called from the general use thereof by all persons concerned in trade and business.

All questions in this rule are performed by taking aliquot or even parts by which means many tedious reductions are avoided. The table of which is as follows.

of a Pound.		of a Shilling		of a Ton.		of an Hundred.	
s d				cwt		qrs lb	
10 " 0 is	$\frac{1}{2}$	6 is	$\frac{1}{2}$	10 is	$\frac{1}{2}$	2 or 56 is	$\frac{1}{2}$
6 " 8 "	$\frac{1}{3}$	4 "	$\frac{1}{3}$	5 "	$\frac{1}{4}$	1 " 28 "	$\frac{1}{4}$
5 " 0 "	$\frac{1}{4}$	3 "	$\frac{1}{4}$	4 "	$\frac{1}{5}$	14 "	$\frac{1}{8}$
4 " 0 "	$\frac{1}{5}$	2 "	$\frac{1}{6}$	$2\frac{1}{2}$ "	$\frac{1}{8}$	of a quarter,	
3 " 4 "	$\frac{1}{6}$	$1\frac{1}{2}$ "	$\frac{1}{8}$	2 "	$\frac{1}{10}$	14 lb is	$\frac{1}{2}$
2 " 6 "	$\frac{1}{8}$	1 "	$\frac{1}{12}$			7 " "	$\frac{1}{4}$
2 " 0 "	$\frac{1}{10}$					4 " "	$\frac{1}{7}$
1 " 8 "	$\frac{1}{12}$					$3\frac{1}{2}$ " "	$\frac{1}{8}$

图7　安·威特曼的练习册最后几页中的一页，日期为1837年4月12日

校的伊丽莎白·道森在“三的法则”上做了很多功课，在“练习”上下了更多的功夫。比如，为了求每码单价6先令8便士对应7234码的值，她用了1971年前所有英国小学生都知道的一个换算比例，即6先令8便士等于$\frac{1}{3}$英镑。不过对于她来说，用每英

尺3先令$7\frac{3}{4}$便士的单价，来求65英尺$3\frac{1}{4}$英寸的费用却并不太容易。

伊莎贝拉·伦德是学生，就读于兰开夏郡博尔顿勒桑德的
65 一所文法学校（该校最初仅面向男孩）。从1866年4月开始，她仅用一年时间便从“简单加法”学到了“三的法则”。一年后，在格洛斯特的里伯斯顿·哈尔女子文法学校就读的G. 琼斯小姐，一张又一张地填写了共20张包裹单。她的8双精纺毛料长袜于1868年7月到了一位叫詹金斯的小姐手里。

以上所选的那些女孩的抄写本，是一些我们正好知道作者名姓、学校及日期的抄本。没有进一步的研究，我们无法假设这些抄本具有代表性。但它们确实显示出对女孩们的数学教育高度重视实践（于此没有欧几里得几何）。另外，就现代的标准来看，教学进度有时极其缓慢而又重复。尽管如此，记录这些抄本的女孩既识字又会算术，尤其是与她们前几代的同龄人相比。

不过，要超越基础数学而进入大学教育，仍然需要具备特殊的才能。我们将比较两位的确曾设法达到其本国教育体系较高水准的女性，来结束这一节的内容。一位是苏格兰的弗洛拉·菲利普，另一位是希腊的弗洛朗蒂亚·丰图科里。

弗洛拉·菲利普是1893年爱丁堡大学第一批女毕业生之一，不过她早在七年前就已经加入了爱丁堡数学学会。实际上，她大部分高等数学方面的教育并非来自大学，而是来自爱丁堡女性大学教育协会。该协会成立于1867年，旨在为女性提供超越学校水平的教育，即与面向男性的大学教育持平。有人认为数学“完全超出了女士的思想范围”。尽管有这些人的反对，数学还是早早就被纳入了协会的授课课程中。这样做的目的是教授与大学相

同的数学，但难度却没有大学课程那么高，因为许多女性在基础
阶段准备不足。尽管如此，她们还是学了欧几里得几何、代数、三 66
角学及圆锥曲线。参加课程的女性通常很少，但正如一位讲师所报告的那样，“班级里的热情和勤奋加倍弥补了人数上的不足”。后来，协会开设了一门更有难度的高级课程。弗洛拉·菲利普于1886年顺利完成了这门课程。就在这一年，她加入了爱丁堡数学学会。1893年，她从大学获得了学位。此时，她已经在协会创立的圣乔治女子学校任教了一段时间。然而，就在同一年她结婚了，从此退出了学术生活，也退出了爱丁堡数学学会。

1869年，弗洛朗蒂亚·丰图科里在雅典出生。她跟弗洛拉的职业轨迹在许多方面都相似。弗洛拉于19世纪80年代在爱丁堡协会学习数学时，弗洛朗蒂亚·丰图科里从雅典的阿塞克翁女子师范学校获得了教师证书。之后，学校的管理委员会资助弗洛拉在柏林学了一年教育学。她后来请求延期，以便在苏黎世获得一个数学学位，但委员会拒绝了。（而另一方面，她的兄弟米海尔确实成了一位数学工作者，后来在汉堡工作。）弗洛拉在圣乔治教书的同一年，弗洛朗蒂亚回到了科孚岛的阿塞克翁学校任教。1892年，就在弗洛拉入读爱丁堡大学时，弗洛朗蒂亚成了雅典大学数学系的学生，她是这里的第一位女性。但与弗洛拉不同，她似乎没有毕业。相反，她继续在雅典的一所女子学校教书，这是她和朋友伊琳·普里纳里自己创办并打理的学校。到了1899年，她签名时总是用丰图科里-斯皮内利，这暗示她可能已与另一位老师卢多维克斯·斯皮内利结婚，但实情却不清楚。不幸的是，在19世纪90年代末，她还未到三十岁，健康状况
就开始恶化。随后她去了意大利生活，于1915年在那里去世。 67

弗洛拉同弗洛朗蒂亚不得不努力来获得她们想要的那种教育。尽管如此，爱丁堡大学和雅典大学在这方面还是走在了其他大学前面。而在剑桥大学，女性直到1947年才被接纳为正式成员。

自学者

直到两个世纪前，世界上还只有极少数的女孩受过某种形式的数学教育。但即便对男孩来说，数学方面的义务教育也是一个相对较新的现象。在17世纪的英格兰，正如我们从沃利斯和佩皮斯的经历所了解到的，完全不学数学而接受从中小学直到大学的教育是可能的。因此，那些对这一学科有特殊天赋或喜好的人通常都是自学的。费马就是这种情况，他在波尔多有位叫艾蒂安·德·埃斯帕涅的朋友，他从这位朋友父亲的藏书中学到了他那个年代最高级的数学。作为17世纪最伟大的数学家之一，艾萨克·牛顿也是如此。牛顿或许已经在位于林肯郡格兰瑟姆的文法学校里学了些基础数学，但他17世纪60年代在剑桥当学生时通过自己的阅读学到了更多的东西。许多年后，他向一位朋友讲述了自己是怎么读笛卡尔的《几何》一书的。这部作品恰在这之前的几年得以用拉丁文重新出版。阅读一本标新立异的数学书，很多人都会承认其中有诸多困难；但也许很少有人会效仿牛顿，顽固而自我激励式地坚持下去。

> 他买了笛卡尔的《几何》一书来自习。他读了二三页后就看不懂了，于是便重新开始；接着又读三四页，直到碰见另一处难点，然后再重新开始并进一步往前走；如此反复，直至他自己成为整个领域的大师。

从遗存的牛顿手稿中我们得知，他以类似的方式来处理同时期的其他作品，并且对从中发现的素材进行研究，从而创建远远超越其任何一位前辈所开创的数学。

在17世纪，以及进入18世纪后的某些时候，一个有足够动力的人仍然可以阅读并学习几乎所有可用的数学文献。即便在19世纪初，索菲·热尔曼仍然设法自学一些当时最高级的数学，但她几乎属于能这样进行学习的最后一代。到了20世纪，几乎没有这种可能了，除非是具有超凡数学直觉的天才，就像自学成才的印度数学家拉马努金那样。安德鲁·怀尔斯当然不是自学成才的，他历经多年的正规教育。如今，即便是在数学上最有天赋的人也需要多年的正规教育，来了解一些问题、技巧以及该学科的传统。几乎人人都能解决诸如费马大定理之类前沿问题的那种"业余"数学的时代，早已一去不复返。

尽管如此，作家们不时地会继续为那些设法从另一人的文稿中学习数学的个人虚构生平活动，好来理解并扩展他们的工作。这样的故事，一个是安娜·麦格雷尔笔下的《爱因斯坦夫人》，另一个是戴维·奥伯恩新近的剧作《证明我爱你》。在这两个故事中，女主角都是数学家的女儿（这种情况我们在现实生活里已经见过），她们设法自我教育，以达到她们父亲研究成果那般的高水平。很不幸，现代数学的现实是，这样的成绩如今极不可信。

究竟为什么学数学？

考虑到数世纪间人们投入了大量的精力进行数学教学，那么就此问一句"为什么？"似乎有些不合常理。不过，随着时间流逝，这个问题的答案已经发生了很大的变化。公元前二千纪的苏美尔文本清楚地表明，识字与算术对社会的合理管理至关重要，尽管对“F号房屋”那庭院里狭小长凳上的男孩们来说，这似乎是个有点遥不可及的想法。

两千年后，在13世纪意大利的算盘学校里接受教育的男孩们，像过去巴比伦的那些同龄人一样，学习处理数字、重量和度量，只是出于不同的原因：并非为了整个社会的利益，而是为了使自己作为个体更适合人们期望他们从事的商业冒险。罗伯特·雷科德为《通往知识之路》一书作了序，还为需要几何知识的特定手工艺和职业列了份长长的名录，序言及名录再次体现了数学技能对于个人的价值。

然而，在雷科德的著作中，我们也瞥见了学习数学的另一个原因：激发智慧，也就是锻炼思维能力。雷科德并不是第一个提出如此建议的人。一些数学难题归在了公元8世纪的那位老师阿尔库因名下，而被称为“阿尔库因用于考验年轻人的命题”。此后，一直存在着应当学习数学来提高思维能力这样的观点，就像学习拉丁语或希腊语那样。毕竟，日常生活所需的数学，比如基本的计时与计账，大多数人在童年的末期就已经掌握了。很少有成年人要用到毕达哥拉斯定理，或者需要求解二次方程以
70 及把一个角二等分，但几乎所有人都曾经学过这些。人们可以（而且我也会）争论说，学习一门外语或者研究历史同样可以促进记忆力以及推理与分析能力的开发，但这些学科从未有过数学所获得的声誉，并且眼下它们在英国的学校课程中只是选修科目，而不是必修。

也许正是因为数学这种纯粹的持久生命力，才使它成为面

向每一个现代儿童的教育中不可或缺的一部分。还有一种情况，即那些想触及该学科前沿的人，需要从年轻时就起步并定期练习，就像年轻的音乐人那样。 71

第五章

靠数学谋生

任何想开辟新天地的数学家都需要时间来思考，来涂涂写写，也需要某种形式的经济支持。不妨让我们暂且来回顾下在第一章中遇到的那些人。我们不清楚丢番图是如何谋生的，也许他靠教书过活，就像许多有数学才能的人那样。在费马之前的时代，许多享有盛名的数学家都曾教过数学，但通常只是作为第二职业：吉罗拉莫·卡尔达诺和罗伯特·雷科德是医生，尽管雷科德一生中也在造币厂和矿山工作过很久；拉斐尔·邦贝利跟西蒙·史蒂文都曾从事实际的建筑项目；弗朗索瓦·韦达则是位律师、顾问，同费马一样。费马经常被描述成“业余”数学家，但在他所生活的时代很少有专业数学家，因此“业余”这个概念毫无意义。而另一方面，我们只能用专业人士来描述怀尔斯。他受到充分认可，且可以从全职的数学研究与教学中获得报酬。

几百年间，数学家受聘用的方式已经发生了重大变化。一个现代的数学家很可能在教育、金融或工业等行业工作，而所有这些工作都是按照制度组织起来的。有些个人可能也准备好了为数学服务、教学课程或许还有会计技能来支付费用，但他们只会雇用极少的工作人员。公元1世纪时，情况却截然不同。欧洲 72
和亚洲大部分地区的经济与政治权力都集中在君主、主教、哈里发和军阀的手中。那些想靠知识方面（包括数学在内）的才能来过活的人，明智地将自己置于一位资助人名下，后者有足够的实力为他们提供资金支持并保护他们。这种资助可以采取多种不同的形式。本章中，我们首先将看到它在三位学者的生活中所发挥的作用。10世纪至11世纪，他们生活在伊斯兰所统治的地域内。

资助方式

塔比·伊本·库拉于公元826年出生在哈兰镇，那里离现在土耳其与叙利亚两国的边界非常近。他早年在那里做货币兑换生意。他不是穆斯林，而属于当地的萨比教派。就在几年前，阿拔斯王朝的哈里发马蒙在巴格达建了座图书馆，它被称作拜特希克玛（意为“智慧宫”），用来保存希腊语、梵文或波斯语的文本并将其翻译成阿拉伯语。巴格达数学家穆罕默德·伊本·穆萨从拜占庭返回时路过哈兰。伊本·库拉在母语叙利亚语之外对希腊语和阿拉伯语的了解，也引起了他的注意。可惜我们不清楚这次会面的日期，但可以推测伊本·库拉当时还比较年轻，因为他应伊本·穆萨的邀请搬到了巴格达。在那儿，他师从伊本·穆萨和他的两个兄弟（世称穆萨家族），学习数学和天文学。

在随后的几年中，伊本·库拉成了巴格达最受尊敬的学者之一。他的文章涉及医学、哲学和宗教，不过如今他却因在数学 73

与天文学方面的工作而广为人知。他把阿基米德的若干论著译成了阿拉伯文，还广泛地就阿基米德所感兴趣的主题写了不少文章，比如力学，以及与弯曲形体的面积、曲面或体积相关的问题。他对托勒密的《天文学大成》进行评注，并撰写关于球形几何和天文学的文章，尤其涉及了太阳的运动与视界高度、月球与当时已知的五个行星的运动。他还深入研究了欧几里得的《几何原本》，并试着证明欧几里得公设中关于平行线的那一条。而在17世纪的牛津，人们再次尝试证明这条公设。伊本·库拉也给出了他自己关于毕达哥拉斯定理的证明，其中一个证明如图8所示。

伊本·库拉一直待在巴格达，直到公元901年去世。他跟穆萨兄弟的情谊保持了许多年，而且还教过伊本·穆萨的儿子。他在生命的最后十年中成了哈里发穆塔迪德的宫廷常客。根据12世纪时作家基夫提一条传记式的简述，他与哈里发关系亲密，所以被允许“在他所希望的任何时候坐在哈里发面前”。后来，他的儿子锡南和两个孙子凭借各自的能力成了著名的学者。在伊本·库拉为人所知的一生中，我们可能会发现两个关键特点。一个是在朋友及家人之间建立起来的教学网络，如此一来就把
74 伊本·穆萨的家人与伊本·库拉的家人联系了起来。这种亲密的人际关系，在撰写本书时已经得到多次留意。第二个特点，尤其就伊本·库拉所处的时代与地域来说，是先后由穆萨家族及哈里发本人所提供的庇护和资助。

伊本·库拉去世七十年后，另一位学者阿布·雷汗·比鲁尼（世称“比鲁尼”）在伊斯兰统治地域的另一端出生，来到了一个不太稳定的世界。他出生的那座村镇在阿姆河（又称奥

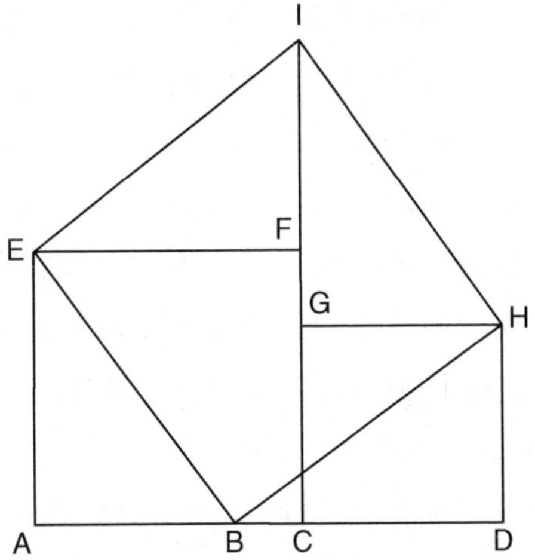

图8　塔比·伊本·库拉关于毕达哥拉斯定理的证明：一个借助简单分割与粘贴的论证，表明$IHBE = EFCA + GHDC$

克苏斯河）畔，位于今天乌兹别克斯坦境内，如今就被称作比鲁尼。他师从数学家、天文学家阿布·纳斯尔·曼苏尔，此后便一直跟随老师工作。他年轻时已开始通过观测太阳来计算当地城镇的纬度，但这种研究活动在公元995年内战爆发时被打断了，而他也被迫逃离。从他对日食及月食的精确观测，我们可以了解他在接下来三十年中广泛的活动轨迹。他有时在里海以南的地区工作，那里靠近现今的德黑兰。据悉他曾把一篇关于编年的著作呈献给了齐亚尔王朝在该地区的统治者伽布斯。其他时间，他在自己的家乡一带生活，起初处于萨曼王朝的统治者曼苏尔二世的庇护之下，后来在阿布·阿拔斯·马蒙的庇护下过了十四年。

最后这段相对稳定的时期于1017年结束了，当时那里被以现今阿富汗东部为中心的伽色尼王朝占据。比鲁尼似乎已被俘虏，后来在喀布尔附近或往南约一百公里的加兹尼住了许多年。他同苏丹马哈茂德的关系尚不清楚：他抱怨受到了虐待，但后来他在进行一些研究时又得到了支持。他还能前往当时同属伽色尼王朝治下的印度北部，并撰写了大量关于这一地区及其宗教、习俗和地理的文章。1030年，马哈茂德去世后，其子马苏德成了
75 比鲁尼在伽色尼王朝的第二位庇护人。马苏德于1040年被谋杀后，其子马杜德成了他的第三位庇护人。比鲁尼本人于1050年在加兹尼去世。

在因朝代更迭而受困扰的一生中，比鲁尼是一位坚定的学者和多产的作家。他的作品约有一半是关于天文学和占星术的，其他则涉及数学、地理、医学、历史和文学。不幸的是，他的作品中只有一小部分留存了下来。

我们要介绍的第三位数学领域的学者，是奥马尔·伊本·易卜拉欣·尼萨布里·海亚米，他在西方更广人知的名字是奥马尔·海亚姆。比鲁尼去世前不久，他出生在伊朗东北部的尼沙普尔。他的名字表明他来自一个制作帐篷的家庭。此时，伊朗地区已经处在源自突厥的塞尔柱王朝统治之下。海亚米年轻时往东到了撒马尔罕，在那里写了一篇关于等式的重要论文，并呈献给了首席法官阿布·塔希尔。后来他在伊斯法罕度过了很多年，在苏丹马里克沙及维齐尔尼扎姆·莫尔克的资助下，他在那里监造天文台并负责编制历法。与此同时，像之前的伊本·库拉一样，他也对欧几里得的《几何原本》写了评注。不幸的是，在尼扎姆·莫尔克被谋杀，马里克沙去世之后，天文台于1092年被关闭。统治者几经更替后，海亚米最终离开了伊斯法罕。他在大致位于伊斯法罕和撒马尔罕之间的梅尔夫待了

一段时间，最后回到尼沙普尔，并于1131年去世。

我禁不住把他的《鲁拜集》中的一首四行诗收录进来。这不是维多利亚时代爱德华·菲茨杰拉德的译文，而是来自1998年沙赫里亚尔·沙赫里亚里的译本：

> 秘密你我永远不知；
> 而且谜底你不知，我也不知；
> 在迷雾后面有着许多与我们有关的话题，
> 可为什么迷雾消散时，你我都未留下。 76

这三个精简的案例研究尚未讲述中世纪时期伊斯兰诸王朝治下数学实践的全部内容，但至少呈现了几个普遍性的主题。主题之一是，就在几个世纪前，地中海东部周边无论何处都可以找到来自希腊的数学作家，但在希腊本土却很少见，所以那些用阿拉伯语来记述数学的人散布在一个更广阔的区域：从现在的土耳其到如今的阿富汗，尽管不是在阿拉伯本土上。出于这个原因，历史学家更乐意将这些作家称为“伊斯兰的”，而不是“阿拉伯的”。但正如伊本·库拉的例子所示，并非所有人都是穆斯林，他们的数学著作也与他们的宗教观点无关。尽管如此，他们所有人都生活在由伊斯兰教习俗和文化占主导地位的社会中，因此“伊斯兰的”这个标签可能比其他的都要好。

随之而来的第二个主题是在一个统治者和王朝都迅速更迭的世界里学术研究的不稳定性。对一个有天赋的男孩或年轻人来说，自己的数学能力得到认可并受到培养，已经是一个事关机会与境遇的问题，就像伊本·库拉和比鲁尼所遭遇的那样。此后，其学习或游历的能力可能在很大程度上取决于统治者的青睐以及资金支持，而统治者自己的未来可能还远未得到保障。单就享有或维持来自对立王朝庇护人的关心而言，比鲁尼看起来在这方面尤其出色。尽管存在这些困难，但其中一些学者的产出却是丰富且多样的。那些在天文学与占星术方面有过著述的人，他们的文章可能还涉及球面几何与三角学、欧几里得的《几何原本》或其他希腊著作者的作品，又或与算术和代数、地理、历史、音乐、哲学、宗教或文学有关。

最后，人们或许会问这种格局对资助者来说意味着什么？不同的个案有很大的差异。在伊斯兰社会中，确实没有单一的词汇能用来形容这里被称为“资助”的关系。正如我们从中国及欧洲的情况中已经看到的那样，统治者往往重视数学上娴熟

77 的人，因为他们有计算吉日的能力。某些情况下，统治者也可能希望长期甚至永久受益于他们对此类善举的支持。此外，拥有有才智的人的服务与陪伴既是享乐的源泉，又是声望的象征。

从大约12世纪末开始，学者们就能够更频繁地在受资助的教学机构马德拉萨获得有薪酬的职位，个别统治者的心血来潮或个人偏好所施加的影响因而减少。不过，为了更仔细地研究从受资助向职业工作的转变，我们现在转向年代稍晚一点的英格兰。

从受资助到职业化

在英格兰，从1580年到1620年的这四十年是一个过渡时期，此时资助仍然存在，但我们也可以看出这方面在向公共责任制薪酬职位发展的最初迹象。托马斯·哈里奥特、威廉·奥特

雷德及亨利·布里格斯这三位的职业生涯，说明当时对英格兰有数学天赋的人来说存在一些可能性和机会。

托马斯·哈里奥特生于1560年，1577年至1580年前后在牛津大学学习。他没有获得数学学位（当时还没有学位一说），但可能已经从导师或他自己的阅读中学到了一些东西。更好的证明是他对探险和航海的兴趣，这似乎也是他从牛津大学学到的，可能经由冒险家理查德·哈克卢特的讲座。16世纪80年代，哈里奥特受到沃尔特·雷利的资助，当时后者对美洲潜在的殖民地非常感兴趣。1585年，哈里奥特在由雷利资助的航行中抵达 78
现在的北卡罗来纳海岸。长达一年的探险以失败告终，但这次探险让哈里奥特和朋友约翰·怀特带回了大量有用的信息，以及一些关于该地区人和动植物种群的漂亮图画。不幸的是，他也带回了对烟草的嗜好，这最终使他丧命。

出行前，哈里奥特受聘于雷利，向水手们教授航海技术，不过他所写的文本现在已经不幸丢失。他回国后继续在雷利的资助下生活，起初住在雷利位于爱尔兰（另一处殖民冒险地）的庄园，后来住进了雷利在伦敦的居所——泰晤士河畔的达勒姆宫。哈里奥特正是在达勒姆宫的屋顶上进行了他早期的自由落体实验，比较了金属球和蜡球的下落速度。哈里奥特与雷利的关系一直很密切，直到1618年雷利被执行死刑的那一天：雷利在行刑台上的遗言，如今还以哈里奥特手稿的形式留存在他的个人文稿和数学论文中。不过到了16世纪90年代初，哈里奥特有了第二位资助人，诺森伯兰郡第九代伯爵亨利·珀西。哈里奥特在珀西位于伦敦的居所泰晤士河畔米德尔塞克斯的塞恩宫，或在他的乡间居所萨塞克斯郡佩特沃斯宫度过了余下的三十年。

不幸的是，哈里奥特的两位资助人都没有成功躲过当时政治与宗教上的紧张局势：珀西同雷利一样，在伦敦塔里被关了许多年。尽管如此，珀西还是给哈里奥特提供了一份收入以及从事他所选择的任何研究的自由。哈里奥特从未对航海问题失去兴趣；他后来也转向天文，而且与同时代的伽利略一样使用望远镜观察太阳黑子和月球陨石坑。他通过牛津的一位朋友纳撒尼尔·托波利设法获得了韦达的数学著作（这些著作后来深刻地影响了费马），因此成为世界上最早一批重视并推广正发展于法国且令人兴奋的新数学思想的人之一，当然，他也是第一个这么
79 做的英格兰人。

哈里奥特没有发表他的发现。有了一份稳定的私有收入，他就无须证明自己或以此谋生。他也没有教书，尽管他确实在朋友圈子中讨论过自己的想法。就某种意义来说，哈里奥特的工作没有多少直接影响；他当然没有像后来的伽利略那样引发知识领域的震动。另一方面，他从事所选工作的自由，使他能探索广泛的主题（其中一些甚至相当神秘玄奥），并使他的研究得出了一些重要的结论。现代术语将此称为“蓝天研究”。哈里奥特的工作成果原本可能很容易就会丢失，但幸运的是，他在同时代的人中的声誉是如此之高，以至于他于1621年去世后，他的文稿得以保存，其中一些思想许多年间都继续在他的后继者中流传。在这个意义上，可以说哈里奥特鼓励了数学讨论以及对数学与科学研究的尊重，尽管是以间接的方式，而正是这些特征塑造了半个世纪后初创的皇家学会。事实上，哈里奥特的声望使皇家学会在最初的十年里不止一次地鼓动人们搜寻他遗存的文稿。

威廉·奥特雷德在创造性上不及哈里奥特，但他在某些方面对于日后英国数学的繁荣来说却与哈里奥特同样重要。奥特雷德生于1573年，只比哈里奥特小几岁，却多活了约四十年。从1610年或更早些时候开始，奥特雷德在萨里郡的奥尔伯里做牧师；他后来除了偶尔造访伦敦外似乎再也没有离开过那里。作为一名既教少儿也教成人的数学老师，他逐渐变得有名望起来。他像哈里奥特一样，也有一位贵族资助人，即阿伦德尔伯爵托马斯·霍华德。这位伯爵在西霍斯利的乡间居所距离奥尔伯里只有几英里。奥特雷德教霍华德的儿子威廉，同时也教当地其他的贵族子弟。通过霍华德，他还遇到了这个家族中的一位亲戚——查尔斯·卡文迪许爵士。这位爵士被证明是英国数学在这一时期的一位重要人物。卡文迪许并不特别擅长数学，但出于某种原因而对它着迷。他热心地收集并试着理解最新的书籍 80
和文稿。比如，他在哈里奥特去世后复制了哈里奥特的手抄本全文，尽管他自己也承认“我怀疑我理解不了所有内容”。正是卡文迪许从法国给奥特雷德带去了韦达的作品，就像早前托波利将韦达的作品带给了哈里奥特那样。

也正是卡文迪许鼓励奥特雷德写了第一本教科书，专门给他的学生——十四岁的威廉·霍华德。这本书于1631年初出版，以其简称《数学的钥匙》而闻名，并且不断印行，有了五个拉丁语版本以及两个英文译本。书中内容比较基础，是算术与代数的入门知识。不过，早前雷科德的教科书至那时已有近一个世纪的历史了，所以人们迫切需要一些新鲜的东西。多年内战之后，当牛津大学任命新教授时，碰巧两位教授都是奥特雷德的学生或读者。他们立即将《数学的钥匙》一书引入牛津，这是这所大学付印出版的第一本数学书。几乎每一位17世纪时有名的数学家，以及许多不知名的数学家，都由这本书迈出了他们的第一步。他们之中有克里斯托弗·雷恩、罗伯特·胡克和艾萨克·牛顿。因此，尽管奥特雷德本人在数学上从未取得过任何重大的进展，而且仅在相对基础的水平上进行教学，但他像哈里奥特一样，在现代英国的早期间接地激励了数学专业知识的传播与发展。

然而，假如没有亨利·珀西、托马斯·霍华德和查尔斯·卡文迪许这三位贵族的鼓励与支持，那么哈里奥特和奥特雷德都做不了他们所做的事情。卡文迪许家族的一位后人在剑桥有一间以自己的名字命名的实验室，而珀西家族和霍华德家族则通
81 常与科学或数学没什么关联。尽管如此，如果没有这三人的信任以及在知识与资金方面的支持，那么这个具有相当规模的数学界于17世纪上半叶在英格兰的兴起可能还需要比实际更长的时间。

同时，与上述事例相比，我们不应该忽视同时代的某些其他发展。1597年，商人、金融家托马斯·格雷欣留下的一笔遗产，被用来开设了天文学、几何学、医学、法学、神学、修辞学以及音乐这七个公开讲席（一周七天，一天一个学科）。格雷欣学院（延续至今且仍在提供公开讲座）在壮大伦敦数学界方面发挥了自己的作用；而17世纪50年代的讲座之后所召开的那些会议，帮助推动了若干年后皇家学会的建立。格雷欣学院的讲席开设二十年后，亨利·萨维尔在牛津大学创立了几何学教席与天文学教席。多年间，格雷欣和牛津之间常有教职人员流动，尤其是格雷欣学院的第一位几何学教授亨利·布里格斯后来也成了牛

津大学的第一位萨维尔几何学教授。

布里格斯来自约克郡的哈利法克斯，年岁与哈里奥特几乎相当。他于1577年进入剑桥圣约翰学院；同一年，哈里奥特在牛津注册入学。但与哈里奥特不同的是，布里格斯始终在大学里发展职业生涯。他在剑桥先教授医学，后来教授数学，然后于1597年转至格雷欣学院。他在学院工作了二十多年，直至接过牛津大学的萨维尔教席。他在这个职位上一直工作到1630年去世。

布里格斯与哈里奥特构成了一对非常有趣的组合：在这一时期的数学史上，一个令人着迷但悬而未决的问题就是他们二人是否曾经见过。这两人应该是碰过面的。在1600年前后，布 82
里格斯和哈里奥特一样，对航海问题非常感兴趣。1610年，哈里奥特在观测太阳黑子，布里格斯则在观测日食、月食。1614年，约翰·纳皮尔提出了对数这项“奇妙的发明”，此后哈里奥特和布里格斯很快就都知道了。布里格斯立即前往苏格兰拜访纳皮尔，并帮助后者进一步发展自己的工作；哈里奥特不再长途旅行，毕竟他已经重病在身，不过，他就对数做着笔记，而且几乎肯定意识到了对数与他自己早期的许多工作有着紧密关联。

人们不禁会想，布里格斯可能跟哈里奥特有过富于成果的长谈，就像同纳皮尔对话那样。这样的场面之所以可能容易出现，是因为在哈里奥特生命的最后二十年中，他俩的居所相距不远：哈里奥特住在塞恩宫，布里格斯住在主教门附近，距伦敦塔都仅有一英里；而哈里奥特正好定期在伦敦塔同雷利和珀西会面。然而，没有证据表明他们的生活曾经有所关联。他们的朋友圈子和影响范围完全不同：布里格斯受雇于一家公共机构，而哈里奥特则在家里为自己工作。1622年，也就是哈里奥特去世后的第二年，布里格斯发表了论文《经由弗吉尼亚大陆通往南海的西北航道》。这篇论文肯定会引起哈里奥特的兴趣。而布里格斯的《对数的算术》直到1624年才发表。17世纪20年代，布里格斯确实与哈里奥特的朋友纳撒尼尔·托波利有过接触，并且得知了为出版哈里奥特一些论文所做的尝试。不过布里格斯本人于1630年去世，也就是哈里奥特的遗作《应用》得以出版的前一年。因此，在发表作品方面，就像生活中那样，他们二人的航迹紧密相连，但冥冥中却又失之交臂。

哈里奥特与布里格斯的生活，呈现了在更有年头的资助习俗和专业数学家的新式生活之间鲜明的对比。后者从明确的职
83 责获得适当的回报，尤其是在教学方面。当然，后者是未来之路。

机构、出版物与会议

约瑟夫-路易·拉格朗日是18世纪最杰出的数学家之一。他的一生概括了在布里格斯和哈里奥特过世一百五十年之后，西欧一位才华横溢的数学家所面对的新机遇。1736年，拉格朗日出生在都灵的一个法意混血家庭（他的洗礼名叫朱塞佩·洛多维科·拉格朗日亚）。他十七岁那年发现了自己对数学的偏爱，两年后便受命在都灵皇家炮兵学校任教。此时，拉格朗日还是同家人一起住在家乡，但在心智上已经开始向往更远的地方。拉格朗日在赴任前不久，将自己的一些工作成果寄给了柏林的莱昂哈德·欧拉。欧拉此时是普鲁士皇家科学院数学部的主管。拉格朗日后来又给欧拉寄去更多的信件，这让他很快当选为科学院的外籍会员。同时，拉格朗日和另一些人在都灵成立

了他们自己的科学学会。这是18世纪50年代在西欧城市得以成立的众多类似学会中的一个，也是今天都灵科学院的前身。

科学协会与学术机构的兴起是18世纪思想文化史的典型特征之一。伦敦皇家学会成立于1660年，巴黎科学院成立于1699年。接着在1700年，普鲁士科学院成立，并于1740年改组为柏林皇家科学院；而圣彼得堡科学院则在1724年以巴黎模式成立。这些机构为少数数学家及科学家提供了职位；更重要的是，他们的例行会议为展示并讨论新兴研究提供了一个论坛。提交给此类会议的论文，随后会在学术性的学报或论文集刊上发表。这个过程可能要花些时间，但论文合集最终都将寄送给遍及欧洲的读者；而且有许多重要的交流通过学术期刊的版面得以进行。拉格朗日将他早期大部分的研究成果发表在《都灵 84
科学论丛》上，这是他自己所在的都灵学会的刊物。

巴黎科学院还建立了为解答问题提供奖励的传统，回应期限为两年。拉格朗日两次参赛获奖，一次是在1764年（解释了“为什么月球总是以同一面朝向地球”），另一次是在1765年（就回答“木星卫星的运动”问题而获奖）。因此到这时，他已经为欧洲著名的数学家们所熟知并受到他们的尊重。比如，早前曾担任《百科全书》科学编辑的让·勒朗·达朗贝尔，就努力为他寻求在都灵以外的职位。1766年，欧拉离开柏林前往圣彼得堡科学院，并提议为拉格朗日在俄国争取一席之地，但拉格朗日反而接受了欧拉在柏林科学院的旧职位。

欧拉与拉格朗日的长期关系始于拉格朗日二十岁之前，所以两人之间既密切又有距离。作为18世纪最多产的数学家，欧拉提出了一个又一个出色而直观的想法，但他在着手处理下一个引发其想象的问题之前，并不总会在这些想法上逗留足够长的时间来研究。经常跟随其后并将半成品的想法转变成合理而优美的理论的人，正是拉格朗日。尽管如此，他俩却从未真正见过面。的确，拉格朗日把欧拉看作自己的前辈和上级，而始终对他恭敬有加却保持着距离。他拒绝直接同欧拉竞争1768年的巴黎奖（关于“月球的运动”），尽管他们最终在类似的一个主题上分享了1772年的奖项。拉格朗日在柏林待了二十年，其间，他在《学院集刊》上（用法文）广泛地发表文章。

腓特烈大帝做了大量的工作来支持柏林科学院。国王过世后，拉格朗日再次迁居，这一次他是前往巴黎科学院，并于1787
85 年抵达那里。两年后，法国的每一家机构都因革命而陷入混乱，不过拉格朗日在那几年设法保住了自己的性命和声誉。1795年，这所科学院被关闭并由国家研究院所取代；拉格朗日则当选为部门主席，该部门负责物理与数学科学。与此同时，大革命对训练有素的教师及工程师的迫切需求促使新的机构得以成立，特别是1794年的巴黎综合理工学院和1795年的巴黎高等师范学院。拉格朗日在这两所院校执教。综合理工学院成了19世纪初巴黎最负盛名的教育机构。任何人，只要学过超出中学水平的数学知识，几乎都会熟悉如下人名：拉格朗日、拉普拉斯、勒让德、拉克鲁瓦、傅立叶、安培、泊松以及柯西。所有这些人都曾在综合理工学院创建初期任教或组织过考试。此外，学校以“书册”的形式发布其课程讲义；这些讲义在整个法国被用作教科书，特别是在那些想被录取为学生的人中。

拉格朗日于1813年离世。在其职业生涯的前三分之二阶段，也就是在都灵和柏林两地，他为所在国家的科学院及其各自的期刊、机构做出了贡献，并从中受益。这些院所做了很多工作来促进新研究的创建与传播。在巴黎的最后几年，拉格朗日看到了一种新式机构的兴起。这种机构旨在为最有才能的学生提供高水平的数学和科学训练。与大学不同，巴黎综合理工学院提供一种重点突出且紧密贴合实用的教育，这将使其毕业生有能力巩固大革命的成果及后来拿破仑帝国的利益。

如果机构的历史看起来有些缺乏人情味，那么我们还可以关注贯穿拉格朗日一生的他与他人密切的人际关系，尤其是他同欧拉和达朗贝尔的关系。拉格朗日过世后，他的门生，家族中 86
一位朋友的儿子奥古斯丁·路易·柯西刚刚开始自己漫长的职业生涯，并在日后成为法国数学界的领军人物，直到1857年去世。从17世纪后期的莱布尼茨到伯努利家族，从欧拉到拉格朗日，再到19世纪中叶的柯西，我们可以追溯出西欧数学界中体现个人友谊与协作的不间断的链迹。

至拉格朗日去世时，他早前的居住地柏林正在发生变化。威廉·冯·洪堡于1810年创建了柏林大学，这是一所不只简单传授累积的知识而是鼓励并促进新研究的机构。德国的大学教授可自由决定自己的职位，并由此来确定其院系的方向与重点。研究团队、研讨会及博士训练等制度都于1900年以前创设于德国的大学，如今被世界上每一所大学或多或少地模仿。从这个意义上讲，包括安德鲁·怀尔斯在内的所有致力于学术的数学家都是19世纪德国的产物。

数学研究成果的发表也在发生变化。在17世纪和18世纪，数学文稿的主要发布渠道是学术期刊。第一份印刷版的数学论文出现在1668年的《皇家学会哲学汇刊》上，由当时的学会主席威廉·布龙克尔撰写。那篇论文只有四页，并同关于《化学、医学与解剖学方面的细节》《每年满潮的种类》这两份给编辑的信件以及一些有关新书的杂项布告列在一起。期刊后来就组织得更好了：例如《博学通报》为医学、数学、自然哲学、法律、历史、地理及神学分别开设了单独的版面；但在整个18世纪，在科学期刊上发表的主题十分多样，数学只是其中之一。

第一种专门的数学期刊——《纯粹数学与应用数学年鉴》，由约瑟夫·热尔冈于1810年在法国创刊并担任编辑，因而也被
87 称为“热尔冈期刊”。这里要注意的是，直到那时，在纯粹数学与应用数学之间还没有出现任何正式意义上的区别。“热尔冈期刊”仅发行至1832年；当时有份相同标题的德语期刊，由奥古斯特·克雷勒于1826年创刊。德文版《纯粹数学与应用数学》（也称为“克雷勒期刊”）一直延续至今。同样，约瑟夫·刘维尔于1836年开始编辑法文版的《纯粹数学与应用数学》（又称“刘维尔期刊”），接替了“热尔冈期刊”。自那时起，数学期刊的出版便蓬勃发展且数量越来越多：如今，期刊不再致力于数学整体，而是专注于该学科的各个分支。我所喜欢的一份期刊是《不适定问题与逆问题》，而除此之外还有数百种之多。

专门机构、入学考试、长期训练、专业期刊、专业学会以及定期会议，是包括数学在内的所有现代专业学科的典型标志。在拉格朗日的时代，还没有国际性会议或者哪怕全国性的会议；但如今确实有了，而且至少占用了所有致力于学术的数学家的部分时间。特别是，数学家们随时准备庆祝彼此的重要生日，这是该学科强大的社会凝聚力的另一个标志。

第一届国际数学家大会于1897年在苏黎世召开，有数个欧

洲国家的代表和美国代表出席了会议。1900年，万国博览会在巴黎举行；同一年，也是在巴黎，召开了第二届国际数学家大会。这一届因德国数学家戴维·希尔伯特所作的报告而广为人知。他概述了二十三个问题，并希望数学家们能在新世纪解决这些问题（尽管证明费马大定理并非其中之一）。1900年以后，大会每四年举行一次，第一次世界大战和第二次世界大战期间除外。88
不过，在20世纪20年代，德国、奥地利、匈牙利、土耳其和保加利亚的数学家被排除在外；另一些人反对这个裁决并缺席会议，这引发了关于这些会议“国际性”的争论。

一份主办城市的名录，讲述了会议自身关于数学研究日益全球化的故事。直到20世纪60年代，每一届大会还都是在西欧、加拿大或美国召开的；但1966年的那届在莫斯科举行，1982年的那届则是在华沙。第一个主办会议的亚洲国家是日本，那是在1990年；2002年中国成为主办方，2010年则是印度。怀尔斯在他的家乡剑桥宣布费马大定理的证明时，可以很容易地同最近三届大会的所在地，即北京、马德里或海得拉巴的与会观众交谈。如今，数学不仅是一门高度专业化的学科，而且是一门完全国际化的学科。

眼下，我们已经到达数学金字塔的顶端。这是个互相之间联系紧密的专业人士社区，并且已经与“数学”和“数学家”这两个词汇联系在一起。自学龄孩童往上，都有正常学习数学的人。不过，比起这些群体的人数，这个专业社区太小了，而其中女性的人数甚至更少。人们会困惑，为什么女性占比仍旧如此之低。对于这个问题并没有简单的答案。但我们应该记得，就像在大多数职业领域里那样，规则都是由男性并且都是为男性制定的；或许是某些女性发现金字塔顶端的空气有些稀薄，而同伴并不总是志趣相投。我们是否把精英数学留给精英历史学家，这无关紧要。就像数学本身已历经许多典型事件那样，数学生活同样以多种形式存在着，没有一种比另一种更合理或更
89 正确。

第六章

进一步了解数学

到目前为止，我回避了过多地讨论数学的技术细节，本章中我们也不会做深入探讨。不过，数学史学家不但要融入过去的数学文本背后的社会环境，而且要尽可能地融入其内容。但这说起来容易做起来难。在某个层面上，过去的数学与如今对大学生的数学要求相比，可能看起来很容易。对历史学家来说，困难通常不是去了解数学本身，而是进入某人的思想和数学世界，不过此人是来自过去的某个时代。

作为示例，让我们考察一下毕达哥拉斯定理，这个定理至此已在本书中被多次提及。欧几里得对该定理的证明如图9所示。这个证明需要在直角三角形的三条边上画正方形，并将最大的正方形分为两部分，接着要说明其中每一部分的大小分别与两个较小的正方形的大小相同。1847年，奥利弗·伯恩巧妙地用彩图展示了证明细节；如图10所示，那几乎是无言的证明。该 90
证明主要特征之一是其适用于任何直角三角形，无论你如何作画（甚至，戴维·乔伊斯的交互式版本将允许您随意地推拉原来

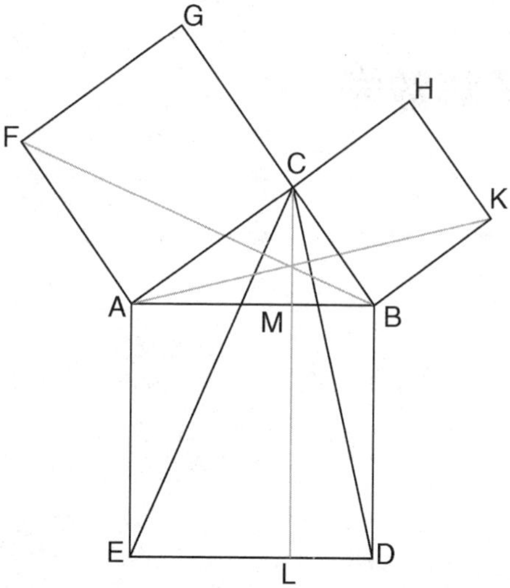

图9　欧几里得对毕达哥拉斯定理的证明：*AFGC* = *AMLE* 且 *CHKB* = *BDLM*

的三角形，只要你固定住其直角）。换句话说，这个证明不依赖于特定的尺寸。其中没有算术，当然也没有代数。这完全符合《几何原本》的风格：欧几里得容许读者用直线和圆规，但不包括辅助计算工具。

塔比·伊本·库拉的证明，如图8所示，同样依赖于经过剪切与粘贴的几何形状，从而说明可以构造较大的正方形来覆盖两个较小的正方形。对欧几里得和伊本·库拉而言，定理及其证明背后的基本直觉都是几何的。

现在来考察一下现代的标记方式，也就是将三角形的边记作 a、b、c，接着写下 $a^2 = b^2 + c^2$。这是否代表了欧几里得心中的
91 定理呢？某种意义上来说，是的。我们知道，边长为 a 的正方形，其面积为 a^2，所以这个公式恰恰是一种非常简洁的方法，用于对

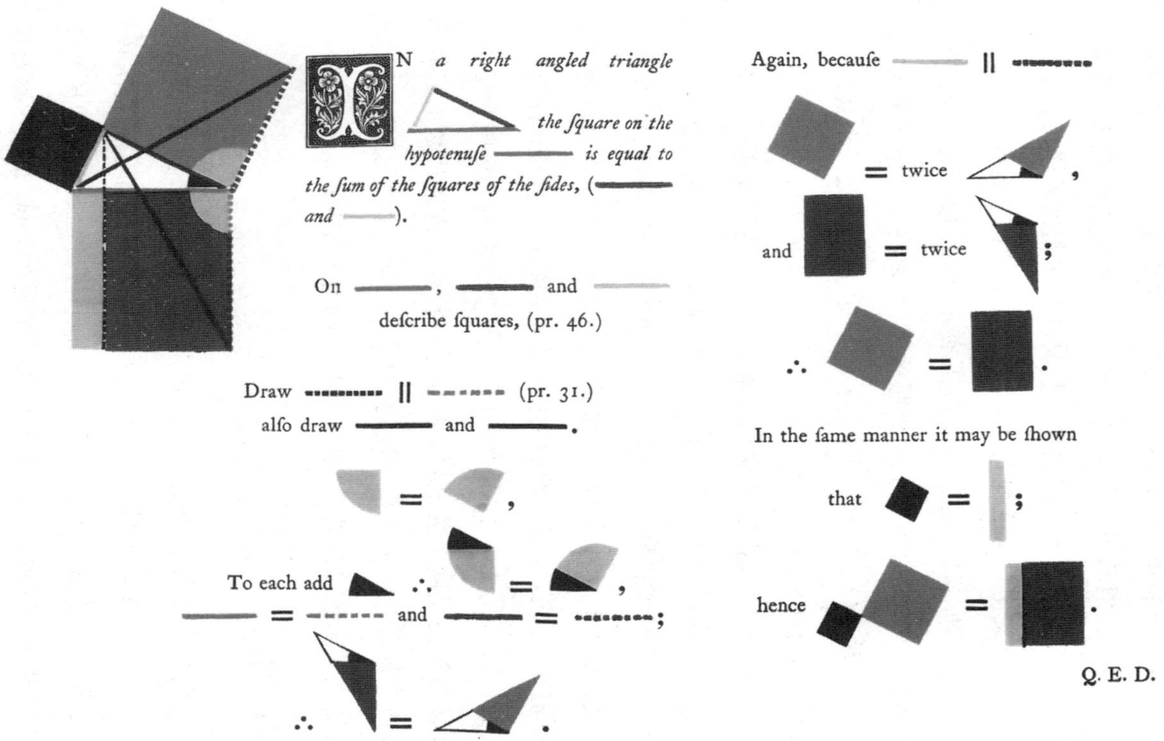

图10　奥利弗·伯恩对毕达哥拉斯定理的证明

一个几何事实进行概述。甚至在语言中也有连续性：我们用同一个词汇“square”来表示a^2这个量以及正方形这种有四条边的
92 几何形状。但从另一种意义上来看，它就无法代表欧几里得心中的定理了。这个公式出自一种与欧几里得的数学文化截然不同的数学文化。在这种文化中，我们学会了让字母代表长度；在这种文化中，我们甚至可以忘记几何形状，而根据其自身的规则来运用字母。因此，我们如果愿意，就可以将上述公式重新写作$c^2=a^2-b^2=(a-b)(a+b)$。这是正确的，但这跟一个直角三角形已不再有明显的相关性。

这种由对几何的洞察到代数运用的转变并非不值一提：学习如何做到这一点需要人们花费不少气力。从历史上看，自几何占主导地位的数学文化转向代数语言开始取得优先地位的数学文化的变化出现在17世纪的西欧。（费马是最早尝试这种可能性的数学家之一，尽管他后来也强烈地抱怨任何偏离传统行事方式的行为。）历史学家深入地研究了这一时期，因为这些变化对于现代数学的发展至关重要。将毕达哥拉斯定理的代数形式看作本质上与其几何形式相一致，就忽略了二者之间的历史鸿沟；而这是众多独立的思想家积累了一代代的努力才跨过的一道鸿沟。

重新阐释

我们刚才看到的是一个在数学方面重新阐释的实例，其中用到了一个几何定理，接着用代数方式对其进行了重新阐释。这是数学家常做的事情。确实，找出一份自己或他人的早期作品，对其进行探究、扩展，在新的条件下进行尝试，是数学家发展其学科的主要方法之一。然而，重写以前的数学对数学家自己

而言是一回事，对历史学家们来说却完全是另一回事。文艺复 93
兴时期，丢番图的《算术》一书在欧洲被重新发现，人们意识到书中包含如此丰富的问题，以至于在数学上和历史上都有多种方式来重新阐释这本书。我们先来看一些数学方面的。

我们已经看到，费马是如何扩展书中II.8这一问题的，并就此在三次方甚至更高次幂上进行尝试，就像对平方那样。下面我们将看到17世纪初对丢番图书中另一个问题的另一次重新阐释，这一次是英国数学家约翰·佩尔提出。1611年，佩尔出生在萨塞克斯郡的索斯威克，他与（我们在第五章介绍过的）哈里奥特和奥特雷德生活在同一时代，尽管他比这两人年轻了四五十岁。他先在索斯威克以北几英里处新成立的斯泰宁文法学校就读，后来在剑桥的三一学院接受教育。之后他回到了萨塞克斯并在奇切斯特一所实验学校任教，直到几年后学校关闭。随后，佩尔花了几年时间寻找有薪酬的职位或资助人，但发现二者都不适合他多少有些独特的性情。1643年底，他被委派到阿姆斯特丹文理中学教书；两年后，他受任命执教于布雷达的伊吕斯特学校，并在那里一直工作到1652年。

在此期间，佩尔对丢番图给予了极大的关注。我们之所以知道这一点，是因为17世纪40年代初佩尔结识了查尔斯·卡文迪许爵士（这位我们在第五章也介绍过），而佩尔在荷兰的那些年他们都有往来。他们基于数学的资助关系有自己的形式：卡文迪许阅读数学文稿时，遇到新近任何未能理解的内容，便会请佩尔帮忙；而佩尔则会及时做出回应。卡文迪许显然非常认可佩尔的能力，并期望佩尔出版一些重要的著作，包括就丢番图的《算术》出版新的版本；对此他写道“我实在太想看了”。可惜

的是，佩尔未能完成或出版任何东西。这几乎不合情理，不过有
94 证据表明他至少已开始着手准备这样一个版本。

证据出自佩尔的大量笔记（共有数千页，如今装订成三十余卷并收藏在大英图书馆内）。佩尔对丢番图如此感兴趣的原因是，佩尔发展了一套他认为非常适合《算术》中那些问题的解答方法。该方法是这样的：首先，对于任何问题，在编号行中记下未知量和给定条件；然后，从条件出发进行系统的演算，得出所要求的解答。为了确保演算正常进行，会分出三列，中间窄的那一列用来标行号。对每一行，左边一列包含一个简短说明；右边一列则显示推演结果。整个过程的观感都与现代计算机算法十分相似。

要了解这种方法是怎样应用于古代丢番图的作品的，不妨让我们来看一下佩尔对《算术》中IV.1这一问题的重现：求解两个数，它们的和是一个给定的数，而它们的立方和为另一个给定的数。丢番图提示说，两个数之和为10，而其立方和应是370。这恰好是后来年轻的安妮·戴夫南特在她父亲指导下演算的问题。佩尔则以自己独特的风格解决了这个问题。他的前两行如下所示，其中他把未知数记作a和b。

$$\begin{array}{r|c|l} a=? & 1 & aaa+bbb=370 \\ b=? & 2 & a+b=10 \end{array}$$

接着，佩尔就像丢番图那样引入了第三个数c，并假设$a=c+5$，因此必然有$b=5-c$。这样，下两行便是

$$\begin{array}{r|c|l} c=? & 3 & \text{令}\, a=c+5 \\ 2'-3' & 4 & b=5-c \end{array}$$

其中，$2'-3'$只是表示用第二行减去第三行。现在一切就绪，演算可以继续进行了。想了解细节的读者需要明白，佩尔的 95
说明$3'@3$表示取第3行的立方，而$10'\omega 2$则是说取第10行的平方根。佩尔偏爱的另一条惯例是，一旦找到所需的值，那么小写字母就换成大写字母。

$3'@3$	5	$aaa = ccc + 15cc + 75c + 125$
$4'@3$	6	$bbb = 125 - 75c + 15cc - ccc$
$5' + 6'$	7	$aaa + bbb = 30cc + 250$
$7', 1'$	8	$30cc + 250 = 370$
$8' - 250$	9	$30cc = 120$
$9' \div 30$	10	$cc = 4$
$10'\omega 2$	11	$c = 2$
$11' + 5$	12	$c + 5 = 7$
$5 - 11'$	13	$5 - c = 3$
$3', 12'$	14	$A = 7$
$4', 13'$	15	$B = 3$

最后四行用来检查这个问题确实已经正确解出：

$14'@3$	16	$AAA = 343$
$15'@3$	17	$BBB = 27$
$16' + 17'$	18	$AAA + BBB = 370$
$14' + 15'$	19	$A + B = 10$

看起来佩尔计划以这种风格重写《算术》的全部六册内容；但如果他全部完成了，那么如今也看不到他的手稿了。不过，他

的方法给同时代的许多人留下了深刻的印象。对此，他的朋友约翰·奥布里甚至发明了一个新的拉丁语动词“pelliare”，意为“佩尔化”。

从上述题例明显可以看出，佩尔不乐意用多余的词汇：19行演算中出现的唯一一处文字是“令”字（实际上他写的是拉丁文“sit”一词）。但如果不用词汇表述，就必须有符号来替代，而佩尔在这方面正是位发明大师。符号“@”和“ω”让左侧一列保持简洁，它们早已不再被使用；不过我们仍在使用佩尔的
96 除法符号“÷”。发明记号是佩尔的特殊才能之一；在这方面，他遵循了当时的一项英国传统中的某些原则。至1557年，罗伯特·雷科德基于两条平行线想出了符号“＝”，“因为除此之外不存在能更进一步相等的两个东西了”。1600年前后，托马斯·哈里奥特新添了不等号“<”和“>”，以及用“*ab*”来表示*a*乘以*b*，这种书写方式如今已成为惯例。1631年，威廉·奥特雷德引入了符号“×”，尽管他很少使用这个符号。他还充满激情地争辩说，符号“把每一步运算及论证的整个方向和过程清晰地呈现在眼前”。显然，这也是佩尔的想法，佩尔的方法使演算过程一目了然，而无须进一步解释。因此，他对丢番图进行“佩尔化”的努力，同17世纪早期英国代数学家们对丢番图及他的《算术》所做的工作相比，更能向我们述说他们的愿望。

立足于历史视角而非数学视角的重新阐释也是如此：比起阐释后的内容，它通常更多地揭示了阐释者本人的信息。例如，数百年间流传的关于代数起源的故事，不仅记录了历史事实，还记录了与阐释同时代的理解。12世纪后期，代数借助于花剌子米所著《还原与平衡》的译本首次来到了西欧的非伊斯兰地

区；不过到了16世纪，即使这段早期的历史曾经为人所知，现在也已经被遗忘。尽管如此，即便只从“algebra”和“almucabala”这两个与之相关却听来奇怪的词汇上，也可以辨识出这一主题源自伊斯兰。因此，16世纪早期的作家们将代数的发明以不同的方式归功于“某个才智超凡的阿拉伯人”，有时则归在另一人的名下（实际指贾比尔·伊本·阿夫拉，12世纪时西班牙的一位穆斯林天文学家，其实他与此事毫无关联），或者又归于“Maumetto di Mose Arabo”这样一个名字含混不清的人（一个阿拉伯人，名字是“穆罕默德·伊本·穆萨”的某种译称）。

不过，德意志学者约翰内斯·穆勒于1462年在威尼斯研究过丢番图的《算术》手稿。他经常又被称作雷格蒙塔努斯，这个 97
名号源自他家乡柯尼斯堡的拉丁化名称。三年后，他在帕多瓦讲演时将手稿的内容描述成“整个算术的精华……这些内容今天我们称之为‘代数’，这个名字来自阿拉伯语”。他的讲演直到1537年才发表。但不久之后，其他作家开始接受相同的主题，即代数是丢番图发明的，只是后来被“阿拉伯人”继承了下来。人们可以看到，这样的故事只因被一个希腊的背景赋予了直接的尊重与地位，就可以让人广泛接受。丢番图所处理的问题，无论风格还是内容都与伊斯兰文本中所发现的那些问题不同，这一事实看起来却并没有妨碍任何人认定后者是以某种方式从前者衍生而来。

即便到了今天，人们对西欧从伊斯兰世界继承下来的数学有了更多了解，丢番图有时却仍被认为是代数的奠基人。这是一场可以一直持续下去的辩论，而我们应该尝试从数学上来理解风险是什么。正如佩尔的例子所展示的那样，丢番图确实提出了若干“求解一个未知数”的问题，这些问题用现代的代数方法很容易处理。不过，他还提出了许多其他“不确定的”问题，也就是说那些问题有多个可能的解答。对于这种情况，丢番图通常是满意的，只要他能通过某种特殊的方法给出其中一个解答即可。实际上，他的工作充满了想法，其中有些还非常聪明，可以用于特定的问题；这跟后来伊斯兰代数文本中通用的规则不同。也有暗示说，丢番图书写时用了某套基本的符号标记，比如将一个未知数记为ς，其平方运算则记为Δ^{Υ}。但现在已经有证据表明，这些分别用来表示希腊语词汇“arithmos”（“数”）和“dynamis”（“乘方”）的缩写，是9世纪的抄录员引入的，根本不
98 能归在丢番图的名下。最后，源自《算术》的数学已经被吸收进现代的数论，而伊斯兰的“al-jabr”（代数）文本更直接地引发了代数研究在西欧的兴起。在我看来，“代数”一词应该保留，由参与者自行将规则与流程描述为“al-jabr”或“algebra”；而对于一位在更早且非常不同的传统中工作的著作者，我们也不应把这些词及其所承载的历史强加在他的身上。

谁是第一个……？

我们刚刚研究了这样一个问题，即“谁发明了代数”。这是数学史学家有时会被问起的一个典型问题，他们经常被期望能够说出是谁第一个发现或发明了某些思想。此类问题极难回答，除非是在最简单的情况下。比如以微积分的发现为例，这是一个可用于描述和预测变化的数学分支，如今，它被用于生物学、医学、经济学、生态学、气象学，以及其他所有与复杂的交互系统协同工作的科学。因此，想要了解“谁发明了微积分”并不

是不合理的。

这一问题的简要回答是，下述两人几乎同时但各自独立发明了微积分：艾萨克·牛顿的工作是在剑桥完成的，而戈特弗里德·威廉·莱布尼茨则是在巴黎。对现代的历史学家来说，就此不再有任何争议。因为我们有两人的手稿，可以清楚地看到他们的思想是何时并以什么轨迹发展的。我们还可以看到，他们以截然不同的方式开展工作，并且发明了各自的表述和符号表示法（莱布尼茨用了“微分”这个词，而牛顿则用了“流数”；莱布尼茨发明了如今人们所熟悉的运算符“dx/dt”，而牛顿用的则是现在不大常见的“$\dot{x}$”）。

不过，他们同时代的人对这个故事根本不清楚。基本事实是，牛顿在1664年和1665年（他二十三岁生日之前）发展了他那一版的微积分学，但后来并没用它来做什么。到17世纪70年 99
代初，牛顿已经就他的光学发现同罗伯特·胡克进行了一场知识分子间的小论争，也许他因此不愿再在微积分上冒险。不管怎么说，他的兴趣至那时已经转移到了炼金术上，接下来的十年他都专注于此。然而在1673年，当时居住在巴黎的莱布尼茨开始独立研究一些同样早已激起牛顿兴趣的问题，并于1684年发表了他第一篇关于微积分的论文，随后在17世纪90年代发表了其他论文。牛顿对此似乎并不在意，可能是认为莱布尼茨的早期工作同他自己已取得的研究成果相比微不足道。不过，牛顿的一些朋友却有着不同的感觉。在17世纪和18世纪之交，他的英国支持者开始暗示，牛顿不仅是相关方面研究的第一人，而且莱布尼茨实际上可能是从牛顿那里窃取了他思想的种子。莱布尼茨1675年在伦敦时曾看过牛顿的一些论文，并在1676年收到过不止一封牛顿的来信。只是他从中学到了什么，以及这与他已取得的发现之间有什么关联，除莱布尼茨本人之外就没有人真正知晓了。

牛顿和莱布尼茨两人从直接冲突抽身而出，但都允许双方好战的追随者通过论证来决一胜负。最终，莱布尼茨于1711年向他所属的皇家学会申诉，对这一论争进行裁决。牛顿作为学会主席，成立了一个几乎不需要开会的委员会，因为牛顿已经在忙于撰写此事的报告了。毫不奇怪，这对牛顿有利。而同样不足为奇的是，这并非该事件的结局：论争一直持续到1716年莱布尼茨离世之后。这番论争也解释了，为什么1809年时坎布里亚郡的英国学生乔治·皮特学的是一门名为“流数”而非“微积分”的科目。

这是一段令人不悦的故事，无人能从中全身而退。重述这个故事的重点意在强调，当时对任何人来说就此事件一探究竟有多么困难：毕竟没有一个人掌握了所有的事实；此外，很难厘清这番论争是关于整个微积分还是关于微积分的某些特殊方面（莱布尼茨就指责英国人在这方面转换了话题）；而且，就像争论的方式一样，被牵连进来的一些分歧从来都不是原先论争的内
100 容。不过，这个故事的另一个重点是，真相的终极证据并非来自当时人们所写或所说的东西——这几乎总是不完整而有所偏向的，而是来自数学手稿本身。

在数学领域，两个人差不多同时提出相似的想法并不少见，就像微积分那样。一旦基础工作得以奠定，一位数学家就可以很容易地像另一位数学家那样进行运用；之后划分起贡献来就变得非常困难，特别是当两人彼此之间有一定的联系时。正是出于这个原因，怀尔斯在研究费马大定理的那些年里非常小

心地把自己封闭起来。就微积分而言，历史学家有足够的文献证据来弄清到底发生了什么，但情况并不总是如此。19世纪初的两位数学家，即布拉格的伯纳德·波尔查诺和巴黎的奥古斯丁·路易·柯西，也发展了一些非常相似的想法，波尔查诺是于1817年，柯西则是在1821年。柯西是否“借鉴”了波尔查诺的想法呢？波尔查诺的作品发表在一份鲜为人知的波希米亚期刊上，不过这对远在巴黎的柯西来说并非难以获得；另一方面，二人都可以以拉格朗日早期的工作为基础独立开展研究。我们还可以从间接证据来评估柯西的工作方式，他常常从别人那里汲取好的想法并对其进一步详加研究。最后，由于缺乏正反两方面的可靠证据，我们根本无法做出评说。

关于谁是某个发现的第一人的另一个问题，可能是要对我们所认为的这一发现实际包含的内容进行定义。例如，在历史上哪个确切的时点，我们可以说出现了“微积分”，而不是一系列最初对牛顿而后对莱布尼茨逐渐产生意义的思想？正如我们 101
已经看到的那样，与查明缔造者们所知的有用事实不同，要精确地指出代数发端于何处，或者毕达哥拉斯定理在哪里被确立为一条形式化定理，是很困难的。几乎所有新近发展的数学都建立在前人工作的基础之上，有时则是以一些卓有贡献的想法为基础。追溯某项特定技术或定理的先驱人物是历史学家的任务之一，不是为了说明谁是第一，而是为了更清晰地了解数学观念如何随着时间推移而变化。

得出正解

欧几里得的系统性演绎风格，作为数学研究方式的黄金标准已达数世纪之久。在此风格中，每个定理均借助于之前已证明的定理和已约定的定义而得以仔细证明。但即便是欧几里得，也不是绝对可靠的。早在公元5世纪，人们关于欧几里得的其中一个公设就提出了问题，而事实证明这些问题很难回答。这条令人烦恼的公设有时被称为“平行公设”。它可以用不同的方式表述；最简单的说法是，假设平面上有一条直线l，直线外有一点P，那么过点P只有一条直线与l平行。对此我们大多数人都不难接受。由此导出的推论是三角形的内角和为180度。这一点我们大多数人也都不难接受。然而，许多欧几里得的评论者认为，平行公设不应该是公理，而应该是定理；也就是说，应该可以通过其他定义和公设以某种方式加以证明。对此进行尝试的人中就有塔比·伊本·库拉和奥马尔·海亚米；牛津的约翰·沃利斯于1663年也做了尝试。之后，一位原本不为人知的数学家，即意大利北部帕维亚的数学教授杰罗拉莫·萨凯里，在1733年尝试了另一种方法。他研究了如果假设三角形的内角和或小于或大于180度，那会发生什么；当然，他希望结论是荒谬的，那样一来这种假设就不成立。不过他错了。假设内角和小
102 于180度的情况让他得到了一些奇怪但始终一致的结果。

一百年后，俄国喀山大学的教授尼古拉·伊万诺维奇·罗巴切夫斯基以及现属罗马尼亚北部的克卢日镇的亚诺什·鲍耶进一步研究了这些想法（又一个差不多同时获得独立发现的例子）；他们都意识到有可能构建一种数学上可接受但绝非欧几里得几何。这个想法让19世纪的思想家们大吃一惊：一个后果是，没有人知道无限空间本身是欧几里得的还是非欧几里得的，就好比我们顺着街道步行时并不能分辨出地球是圆的还是平

的。数学原本应该提供关于世界的无可争辩的真理；但突然之间，这样的真理似乎变得不那么可靠了。

所有这一切的一个后果是，数学家开始更加仔细地研究他们的基本假设。这些假设的正式名称叫“公理”。的确，19世纪末及20世纪初，在回归真正的欧几里得风格的过程中，数学的所有分支都建立在公理基础之上，这样就为它们附上了自希腊时代以来数学所缺失的逻辑上的严密性。在公元前2世纪至公元19世纪之间如此漫长的岁月中，数学在很大程度上是以完全随机的方式发展的。事实是，数学家并非通过建立公理并对其进行逻辑思考来做探究，而是通过如下方式，即对感兴趣的问题做出富有想象力的回应，面向新的方向抛出问题，或者观察明显不同的数学分支如何能以崭新的方式结合起来。当然，他们必须正确地运用自己的技能和经验；最后，他们必须提出被称为“证明”的严密论证，就像怀尔斯在剑桥讲演中所做的那样；但从初始的见解以及随之而来的艰苦工作走到这一步很可能是一段不短的路程。

上一节中，我们讨论了微积分的发现。它根本不是发端于逻辑，这是数学中的一个经典实例。整个想法是以17世纪数学 103
家们所谓的“无穷小量”为基础。然而，关于“无穷小量”不得不问的第一个问题是，它究竟有没有大小？如果有大小，那么它就不可能“无穷小”；但如果没有，那么它甚至就不存在，而且无法以任何合理的方式应用在计算中。这看起来像是在挑剔，更像是在讨论针尖上能站天使这样的哲学命题，而不是在讨论数学。但这很重要，因为“无穷小量”的讨论会迅速地导出矛盾；而且由于数学被认为应该是统一的逻辑大厦，所以哪怕一个单一的矛盾就会使整个大厦垮塌。(正是出于这个原因,如果数学家们要证明某事不成立,通常会像萨凯里那样故意去设立一个矛盾。这一技巧被称为“归谬法”。)

牛顿和莱布尼茨都深知“无穷小”的悖论并各尽全力来处理,牛顿从正面应对,而莱布尼茨则从侧面迂回。继他们之后的那些人,即不单数学家,还有受过良好教育的公众,对此也有所了解。例如,乔治·贝克莱主教在他那本名为《分析学家:致一位异教徒数学家的演讲》的书中问道:“在宗教观点上如此谨慎入微的数学家,在他们自己的科学里是否也完全一丝不苟呢?他们是否不屈从于权威,而是信任别人所叙述的事物并相信不可思议的观点?”这种担忧是否会让数学家们停滞不前?答案是不会,因为在微积分发展的初期,他们已经意识到它有多么强大,并忙于将其应用于光线、悬垂链、自由落体、振动弦以及物理世界的许多其他现象之上,并且取得了巨大的成功。他们几乎不会因为一个被其视作形而上学方面而不是数学上的困难来放弃这一切。大约用了一百五十年时间,这一问题才得以解决至大多数人满意的程度;解决方式有些过于技术性,就不在这里讨论了。然而,同样是在这一百五十年间,数学超越了所有人的期
104 望,尽管其基础并不稳固。

19世纪也有类似的故事。1822年,巴黎综合理工学院的讲师约瑟夫·傅立叶发表了一篇关于热扩散的论文,即他的《热的分析理论》。文中,傅立叶研究了用正弦函数与余弦函数的无穷和来描述周期分布的想法。这样的无穷和如今被称为“傅立叶级数”,在工程及物理领域有广泛的应用。不过,傅立叶的原始推导中满是错误及矛盾之处。其中一些相互抵消了;但有很多

傅立叶忽略的部分，假如他确实曾注意到的话。换言之，傅立叶级数的初始理论并没有比微积分更牢固的基础；不过像微积分一样，它被证明是一个极富变化而实用的工具。但就像对微积分那样，傅立叶之后的许多数学家也不得不花费大量时间来修补漏洞。

这些实例并非例外。正如我们所见，比傅立叶更有能力的数学家怀尔斯，不得不经历一个非常类似的纠错过程，尽管这在他的个例中只要两年，而不是一个世纪。数学中几乎每一个新的发现都是在一个粗略而现成的状态下起步的，必须加以改进并完善才能呈现给同行，更不用说教给初学者了。

大多数现代的教科书都遵循着同欧几里得《几何原本》一
样的模式，从简单的起点开始，以紧密顺畅的逻辑流程来构建数
学。也就是说，我们允许或者期望学生们跟随一条甚至连最初
的探索者都不曾看见的清晰路径。如果学生们有机会回顾那些 105
初始的发现，很可能会找到非常不同的东西：某个反复尝试的过
程，以错误开端而以死胡同收场；半成形的想法，对应着进行到
一半的工作，而留待别人去发展；经年累月才得以改善的较成形
的构想；所有这些最终都得由这样的老师来修正——他们可能
并非创新者，但必须具备同样重要的天赋，即明白如何向初学者
进行解释。换句话说，一本经过润饰的教科书极少向我们谈及
起初走进数学时的那种直觉与艰辛工作、想象力与奋斗。那是
历史学家的任务。 106

第七章

不断发展的数学史

过去的几个世纪中，编撰及思考数学史的方式已经发生了很大的变化；其中有些与更普遍的思想文化史上的变化保持着一致，有些则是数学特有的。正如我们在第二章所见，约翰·利兰在16世纪50年代采用的方法，被一个世纪后的约翰·赫拉德·福西厄斯所沿用，也就是尽可能多地将有关作者、日期和文本的许多事实记录下来，而不对那些文本所包含的内容进行任何分析。不管怎样，至17世纪后期，任何对数学感兴趣的人都很清楚，该学科的力量、范围和技术正在迅速发展："几何每天都在前进"，约瑟夫·格兰维尔于1668年如是写道；而仅仅几年之后，约翰·沃利斯便称颂了使代数达到"现今高度"的"发展与进步"。

18世纪是个"百科全书式"的时代，有两本专注于数学史的重要出版物问世：让·埃蒂安·蒙蒂克拉的《数学史》，1758年在巴黎出版（1799年至1802年扩展为四卷本）；以及查尔斯·赫顿的《数学与哲学词典》，其中收入许多历史性文章，

1795年出版于伦敦。但至19世纪后期，关注焦点发生转移，就像在其他研究领域那样，从二手记载转向了古代及中世纪文本的学术性版本与译本（文艺复兴时期也是如此）。以本书前面 107
所讨论的文本为例：丢番图所著《算术》的第一个英文译本由托马斯·希思出版于1885年；希思版的欧几里得《几何原本》，以当时可获得的最好的学术成果为基础，于1908年问世；查尔斯·路易斯·卡尔平斯基根据中世纪的一个拉丁语版本翻译了花剌子米的《代数》，仅仅几年后于1915年出版。这样的版本无论在过去还是现在都是非常珍贵的：《算术》和《代数》以前都没有英文译本；希思版的《几何原本》直到今天仍是标准的英文版本。

不过，现代历史学家也以谨慎的态度来对待此类版本。希思关于《算术》的论文，以“亚历山大的丢番图：希腊代数的历史研究”为标题，这一标题所提出的问题在本书中已经提及。此外，就阿波罗尼乌斯著作的希思版译本，一位评论员观察到“幸亏这本书进行了有效的缩编并用现代的符号替代了文字证明，这样它所占用的空间还不到原始文本的一半”。这又是个实例，历史学家或许并不会感谢希思的这项技巧，而宁愿看到未缩编的文本，这样一来就不会有现代标记所造成的时代错位。然而，许多20世纪初期数学史方面的学术研究是由数学家而不是由历史学家完成的，且处理方式大致相同，即将最初用埃及象形文字、苏美尔语、梵文或希腊语写成的文本译介成现代数学的符号和概念。译者们的动机本身并不应该受到指责：在试图理解乍一看似乎完全陌生的观念时，把它们同更熟悉的事物联系起来是很自然的；危险在于，对自己不熟悉的想法，人们仅将其视作

我们如今能更有效完成之事的过时体现，就像我们所看到的那样。这样一来，历史就会被重写，但并非遵从原作者的观点，而
108 是从我们自己的视角出发。

在最早对现代化所导致的扭曲进行抵制的人士中，就有研究古代数学的历史学家。他们于20世纪90年代率先尝试，尽可能地恢复并保存原始文献的习语风格及思想过程。雷维尔·内茨编辑并翻译了阿基米德的作品。正如他在一则如今经常被引用的评论中所说的，“学术性翻译的目的，就我所理解的来说，是消除和外语本身有关的所有障碍，而保留所有其他障碍”。这便迫使当今历史性数学文本的读者付出比五十年前的读者更多的努力，但他们因此在历史认知方面的收获也会无比丰硕。

那些研究古代数学著作的人也引领了历史编纂工作的其他方面，部分原因是过去他们的资料就是用那种方式随意而无序地收集而来的。比如，单单一块黏土板能告诉我们的极为有限，除非我们知道它是在哪儿或在什么时候写成的。假如我们要就某份特定文本是如何与在同一地区或别处所发现的其他文本发生关联勾勒出一幅图画，那么上述信息便至关重要。许多早期发掘的黏土板，或被存放于博物馆而关于其来源只有很少的信息，或被售卖于古物市场而根本不知出处，这就使得历史学家们现在很难从中推断出有用的信息。幸运的是，今天的考古学家在移开每一层证据之前都会非常仔细地记录位置和周边事物。对于用笔墨书写而褪色或残损的文本，现代技术还推动了阅读上的进步。第三章中，我们提到了被重新发现的阿基米德的文本。在这方面，对该文本所做的修复工作就尤为出色。学者们不仅能阅读大部分的原始文本，而且还能辨认出那位把羊皮纸

擦拭干净并在上面重新书写的抄录员：约安尼斯·米罗纳斯，1229年的大斋节期间他在君士坦丁堡工作。恢复文本本身和重现文本背后的故事应该齐头并进，这再合适不过。 109

数学史学家已逐渐摆脱了纯粹的“内在论者”的观点。在这种观点中，数学发展被看作是自发产生的，而无论外界影响如何。正如本书中一遍遍所展示的那样，数学活动在长达数世纪的岁月里以多种方式呈现出来，而所有方式都是由社会和文化决定的。不管怎样，我们倒洗澡水时不应该连同婴孩一起倒掉：数学家们常常致力于解决某个特定的问题，不是因为它可能有用或因为有人要求他们那样做，而是因为问题本身引发了他们的想象。牛顿与莱布尼茨发明微积分，鲍耶和罗巴切夫斯基探究非欧几里得几何，或怀尔斯证明费马大定理，恰恰都是这种情况。在这些实例中，进步首先取决于对数学的深入和专注的参与。在这个意义上，数学创造力可说是一个内在自发的过程。不过，那些在特定时间或地点被认为是重要的数学问题，它们出现的方式，它们被理解、被阐释的方式，都受到数学本身以外的众多因素影响，比如社会、政治、经济及文化等诸多方面。对历史学家而言，背景已经变得和内容同样重要。

近年来另一个重大变化是，人们日渐认识到，少数著名数学家所做的数学，并未反映出社会其他层级数学活动与经验的多样性（尽管是建立在此基础之上）。非精英式数学的历史一直是本书的核心主题之一。像许多其他学科的学者一样，数学史学家对性别和种族问题也变得更为敏感。对进入现代西欧之前或以后的文化进行研究，过去由于资料的缺乏或语言上的障碍而受到了制约。但随着网络图片、新式翻译和学术性评论正使 110

越来越多的原始资料无论在知识还是物质层面都更容易获得，上述情况当下正发生变化。因此，过去的数学不再单单被视为现今数学的前身，而是被看作其当代自身文化的一个组成部分。

就像如今所有蓬勃发展的学科一样，投身于数学史研究的人必须要跨越界限。确实，从事该学科工作的最大乐趣之一就是，可以从考古学家、档案管理员、汉学家、古典学家、东方学家、中世纪学家、科学史学家、语言学家、艺术史学家、文学批评家、博物馆策展人以及许多其他专家的专业知识中学习。原始资料的范围也同样扩大了，不再局限于曾经阐述最新思想的书籍或手稿，而是包括书信、日记、笔记草稿、练习册、测量仪器、计算设备、绘画、素描以及小说在内。最后一项听起来可能令人吃惊，但小说家或许是当代数学观点最敏锐而最清晰的记录者；有兴趣跟进此话题的读者，可以在“延伸阅读”一节找到更多的内容。

过去的五十年中，历史学家所提出的问题已经发生变化且多样化起来。只是问谁在何时发现了什么，已经不够了。我们还想知道，什么样的数学实践吸引了一群人或是个别人，以及为什么。有哪些历史或地理影响在起作用？参与者或其他人如何看待数学活动？哪些方面受到特别重视？人们采取了哪些措施来保存或传承数学专业知识？谁为此支付费用？作为个体的某位数学家是如何管理自己的时间和技能的？他们的动机是什么？他们产出了什么？他们用它来做了什么？以及他们这一路上都在同谁讨论、合作或争论？

111 大多数这类问题的大多数答案都很难有任何确定性。数学史学家也像所有其他历史学家一样，研究零散而不严密的证据；

从中他们必须尽可能仔细地重建事关历史却并不完整的故事。这种尝试对于我们可以了解的一项人类活动仍是重要而值得的。它就像产生文学或音乐的活动一样古老而广泛，并且已经以多种文化形式丰富地表现了出来——这项活动就是从事并创建数学。 112

索　引

（条目后的数字为原书页码，
见本书边码）

A

B

D

E

F

G

H

I

J

K

L

M

N

O

P

Q

R

S

T

U

V

Jacqueline Stedall

THE HISTORY OF MATHEMATICS

A Very Short Introduction

Contents

Acknowledgements

In writing this *Very Short Introduction* to a very large topic I have been greatly inspired by other authors in the series, so many of whom have risen to a similarly demanding challenge in imaginative and thought-provoking ways.

It has been my privilege over the last few years to edit both *The Oxford Handbook of the History of Mathematics* and the *BSHM Bulletin*, the journal of the British Society for the History of Mathematics. This has led me into close working relationships with more than eighty authors writing on the history of mathematics from a wide variety of perspectives. I have learned something from every one of them. Much of that work was done alongside Eleanor Robson, best of friends and colleagues, and I am immensely grateful to her for the hours of companionship and discussion that have helped to shape the picture I have attempted to convey in this book. In particular, I have drawn on the research and expertise of Markus Asper, Sonja Brentjes, Christopher Cullen, Marit Hartveit, Annette Imhausen, Kim Plofker, Eleanor Robson, Corinna Rossi, Simon Singh, Polly Thanailaki, and Benjamin Wardhaugh; books and articles by these authors and some others will be found in the suggestions for further reading at the end.

The John Hersee collection of children's copy books, discussed in Chapter 4, is the property of the Mathematical Association, and is housed in the David Wilson Library at the University of Leicester. I thank the Association's archivists, Mary Walmsley and Mike Price, for their generous hospitality and collaboration in that part of my research. I am also grateful to Joanna Parker of Worcester College, Oxford, for allowing me to see John Aubrey's copy of Anne Ettrick's notebook. I am indebted to Andrew Wiles, Christopher Cullen, Eleanor Robson, and Adam Silverstein for taking the trouble to check details in Chapters 1, 2, 4, and 5 respectively. I record my warm thanks to them and all others who have commented astutely on various aspects of the text: anonymous OUP readers, together with Peter Neumann, Harvey Lederman, Jesse Wolfson, and all the members of my immediate family, some of whom never thought to read about the history of mathematics until now.

List of illustrations

Introduction

Mathematics has a history that stretches back for at least 4,000 years and reaches into every civilization and culture. It might be possible, even in an introduction as very short as this book, to outline some key mathematical events and discoveries in roughly chronological order. Indeed, this is probably what most readers will expect. There can be several problems, however, with that kind of exposition.

The first is that such accounts tend to portray a whig version of mathematical history, in which mathematical understanding is generally perceived to progress onwards and upwards towards the splendid achievements of the present day. Unfortunately, those looking for evidence of progress tend to overlook the complexities, lapses, and dead ends that are an inevitable part of any human endeavour, including mathematics; sometimes failure can be as revealing as success. Besides, by defining present-day mathematics as the benchmark against which earlier efforts are to be measured, we can too easily come to regard the contributions of the past as valiant but ultimately outdated efforts. Instead, in looking to see how this or that fact or theorem originated, we need to see discoveries in the context of their own time and place.

A second problem, about which I shall have more to say later, is that chronological accounts all too often follow a 'stepping stone' style, in which discoveries are placed before us one after another without the all-important connections between them. The aim of the historian is not merely to compile dated lists of events but to throw light on the influences and interactions that led to them. This will be a recurring theme of this book.

A third problem is that key events and discoveries come to be associated with key people. Further, in most histories of mathematics, most of those people will have lived in western Europe from about the 16th century onwards and will be male. This does not necessarily reflect Eurocentric or sexist attitudes on the part of the writers. The rapid pace of development of mathematics in the masculine culture of Europe since the Renaissance has led to a large amount of material that historians have rightly thought worth investigating; besides, we have a wealth of sources from Europe for this period, as opposed to only a handful, in relative terms, for pre-medieval Europe, China, India, or America. Fortunately, the availability and accessibility of sources from some of these other areas is beginning to improve. The fact remains, however, that focussing on big discoveries rules out the mathematical experience of most of the human race: women, children, accountants, teachers, engineers, factory workers, and so on, often entire continents and centuries of them. Clearly this will not do. Without denying the value of certain notable breakthroughs (and this book will begin with one of them) there have to be ways of thinking about history in terms of the many who practise mathematics, not just a few.

This book can do only a little to redress the masculine bias of most depictions of the history of mathematics; it can, however, pay more than lip service to the mathematics of continents other than Europe; and it will attempt to explore how, where, and why mathematics has been practised by people whose names will never appear in standard histories. To do so, however, requires something different from the usual chronological survey.

The alternative model that I propose to follow will be built around themes rather than periods. Each chapter will focus on two or three case studies, chosen not because they are in any way comprehensive or exhaustive but in the hope that they will suggest ideas and questions and fresh ways of thinking. At the same time, in keeping with the ideals proclaimed above, I have tried wherever possible to draw out contrasts or similarities between the various stories, so that readers can build up an interconnected view of at least a few aspects of the very long history of mathematics. My aim has been to demonstrate not only how professional historians now approach their discipline but how the layperson too can think about mathematical history.

In this way, I hope that this book will help the reader to recognize the richness and diversity of mathematical activity throughout human history; and that it will be a very short introduction not just to some of the mathematics of the past but to the history of mathematics itself as a modern academic discipline.

Chapter 1
Mathematics: myth and history

It is not often that a thorny old mathematical problem makes the news, but in 1993 newspapers in Britain, France, and the United States announced that a 40-year-old mathematician called Andrew Wiles, in a lecture at the Isaac Newton Institute in Cambridge, had demonstrated a proof of a 350-year-old problem known as Fermat's Last Theorem. As it turned out, the claim was a little premature: Wiles's 200 pages of mathematics contained an error that took a little while to fix, but two years later the proof was secure. The story of Wiles's nine-year struggle with the theorem became the subject of a book and of a television film in which Wiles was moved to tears as he spoke of his final breakthrough.

One reason that this piece of mathematical history so caught the public imagination was undoubtedly the figure of Wiles himself. For seven years before the Cambridge lecture, he had worked in near isolation, devoting himself single-mindedly to the deep and complicated mathematics underlying the theorem. Here then was a story to which those brought up in the mythologies of western culture were already well attuned: the lonely hero struggling against the odds to attain an elusive goal. There was even a princess in the background: only his wife knew of Wiles's ultimate purpose, and was the first to receive the finished proof, as a birthday present.

A second reason is that although the eventual proof of Fermat's Last Theorem was fully understood by perhaps no more than 20 people in the world, the theorem itself is easily stated. Wiles was already intrigued by it when he was 10 years old, and even those who have long ago forgotten most of the mathematics they ever learned can grasp what it is about; we will return to it in a moment.

Before that, however, note that in the very first sentence of this chapter three people were already mentioned by name: Wiles, Newton, and Fermat. In mathematics this is typical: it is universal practice for mathematicians to name theorems, conjectures, or buildings after one of the tribe. This is because most mathematicians are keenly aware that they continuously build on work done by their predecessors or their colleagues. In other words, mathematics is an innately historical subject in which past endeavours are rarely far out of mind. To begin to think about the questions that historians of mathematics ask, let us pursue Fermat's Last Theorem backwards from that Cambridge lecture theatre in 1993 to its more remote beginnings.

Fermat and his theorem

Pierre de Fermat, born in 1601, spent his entire life in southern France. A lawyer by training, he was a Counselor to the Parlement of Toulouse, the judicial body for a large surrounding area. In his spare time, which was little enough, Fermat worked on mathematics, and being far removed from circles of intellectual activity in Paris he did so almost entirely alone. During the 1630s he corresponded with mathematicians further afield, through the Parisian Minim friar Marin Mersenne, but during the 1640s, as the political pressures on him increased, he withdrew once again into mathematical isolation. Fermat achieved some of the most profound results of early 17th-century mathematics but for the most part was prepared to say tantalizingly little about them. Time and again he promised his correspondents that he would fill in the details when he had enough leisure to do so, but that leisure never came. Sometimes he would offer a bare statement of what he had found, or he would send out challenges that plainly demonstrated the ideas he was working on but without giving away his hard-won results.

The first hint of his Last Theorem appeared in such a challenge, sent to the English mathematicians John Wallis and William Brouncker in 1657; they failed to see what he was driving at and dismissed it as being beneath their dignity. Only after Fermat's death, when some of his notes and papers were edited by his son Samuel, did the full statement of the theorem emerge, scribbled in the margin of Fermat's copy of the *Arithmetica* of Diophantus. Before taking another step back in time to see what it was in Diophantus that inspired Fermat, we need to digress briefly to some mathematics, Fermat's Last Theorem itself.

The one bit of mathematics that almost everyone recalls from their schooldays is Pythagoras' Theorem, which states that the square on the longest side of a right-angled triangle, the hypotenuse, is equal to the sum of the squares on the two shorter sides, the 'legs'. Most people will probably also remember that if the two short sides are respectively 3 and 4 units in length then the long side will be 5 units, because $3^2 + 4^2 = 5^2$. This kind of triangle is known as a 3-4-5 triangle and may conveniently be used for marking out right angles on the ground with a piece of rope, or by textbook writers who want to set problems that can be solved without resort to a calculator. There are plenty of other sets of three whole numbers that satisfy the same relationship: it is easy to check that $5^2 + 12^2 = 13^2$, for instance, or that $8^2 + 15^2 = 17^2$. Such sets, sometimes written as (3, 4, 5), (5, 12, 13), and so on, are known as 'Pythagorean triples', and there are infinitely many of them.

Now suppose that, as mathematicians like to do, we tweak the conditions a bit and see what happens. What if, instead of taking the squares of each number, we take their cubes? Can we find triples (a, b, c) that satisfy $a^3 + b^3 = c^3$? Or can we be even wilder and ask for a triple that satisfies $a^7 + b^7 = c^7$ or even $a^{101} + b^{101} = c^{101}$? The conclusion Fermat came to was that there is no point in trying: we cannot do it for any power beyond squares. As so often, however, he left it to others to work out the details. This time, his excuse was not time but space: he had discovered a marvellous proof, he said, but the margin was too meagre to contain it.

The margin in question belonged to page 85 of Claude Gaspar Bachet's 1621 edition of the *Arithmetica* of Diophantus. The *Arithmetica* had intrigued European mathematicians ever since a manuscript copy, written in Greek, had been rediscovered in Venice in 1462. About Diophantus himself, no-one knew anything, and little more is known now. The manuscript refers to him as 'Diophantus of Alexandria' so we may suppose that he lived and worked for a significant part of his life in that Greek-speaking city of northern Egypt. Whether he was a native Egyptian or an incomer from some other part of the Mediterranean world, we do not know. And any estimate for his dates is no more than guesswork. Diophantus cited a definition from Hypsicles (c. 150 BC), while Theon (c. AD 350) cited a result from Diophantus. That pins him down to within 500 years, but we cannot do better than that.

Compared with the geometric texts that have survived from other Greek mathematical writers, the *Arithmetica* is highly unusual. Its subject matter is not geometry but nor is it the arithmetic of everyday accounting. Rather, it is a set of sophisticated problems asking for whole numbers or fractions that must satisfy certain conditions. The eighth problem of the second book, for example, asks the reader to 'divide a square into two squares'. For our present purposes, we may translate this into a more modern mode of expression and see that Diophantus' question was related to Pythagorean triples, where a given square (in the notation above, c^2) may be divided or separated into two smaller squares $(a^2 + b^2)$.

Diophantus showed a clever way of achieving this when the largest square is 16 (in which case the answer involves fractions); and then he moved on to something else.

Fermat, however, hesitated at this point and must have asked himself the obvious question: can the method be extended? Can one 'divide a cube into two cubes'? This was precisely the question he posed to Wallis and Brouncker in 1657 (and to which, after Fermat had later reported that it was impossible, Wallis angrily retorted that such 'negative' questions were absurd). What Fermat suggested in the margin in fact applied not only to cubes but to any higher power at all, a long way beyond anything required by Diophantus.

One other name has recurred throughout the above account, so let us now take one further historical step backwards, from Diophantus to Pythagoras, who is supposed to have lived on the Greek island of Samos around 500 BC. Despite this very early date, many readers will probably feel much more at home with Pythagoras than with Diophantus: indeed the question I am most commonly asked as a historian of mathematics is: 'Do you go all the way back to Pythagoras?' It is true that Pythagoras' Theorem has been known for a very long time; the disappointing news is that there is no evidence to link it to Pythagoras. In fact, there is little evidence to link anything to Pythagoras. If Diophantus is a shadowy figure, Pythagoras is buried under a blanket of myth and legend. We have no texts written by him or his immediate followers. The earliest surviving accounts of his life are from the third century AD, about 800 years after he lived, by writers with their own philosophical axes to grind. His supposed journeys to Babylon, or to Egypt, where he was said to have learned geometry, were probably no more than fictions used by such writers to bolster Pythagoras' standing and authority. As to the tales of what his followers are supposed to have done or believed, there may be some foundation for them in fact, but it is impossible to be certain of any of them. In short, Pythagoras became, literally, a legendary figure, to whom much was ascribed but of whom little was known in reality.

The lives of these four men, Pythagoras, Diophantus, Fermat, Wiles, span more than 2,000 years of mathematical history. We can certainly trace similar mathematical ideas running through the stories about each of them even though they are spaced several centuries apart. Have we then 'done' the history of Fermat's Last Theorem from start to finish? The answer is 'no', and for many reasons. The first is that one task of the historian is to disentangle fiction from fact, and myth from history. This is not to underestimate the value of either fiction or myth: both embody the stories by which societies define and understand themselves, and may have deep and lasting value. The historian, however, must not allow those stories to obscure evidence that may point to other interpretations. In the case of Pythagoras, it is relatively easy to see how and why tales that appear robust have been spun from the flimsiest of threads, but in the case of Andrew Wiles, where we believe we have the facts in front of our eyes, it is much harder. The truth of almost any story is almost always more complex than we first imagine or than the authors would sometimes have us believe, and stories about mathematics and mathematicians are no exception. The rest of this chapter examines some common myths and pitfalls in the history of mathematics; for convenience, I have called them 'Ivory tower history', 'Stepping-stone history', and 'Elite history'. The rest of the book will then offer some alternative approaches.

Ivory tower history

One of the most remarkable features of Wiles's story is the fact that he deliberately shut himself away for seven years so that he could pursue the proof of the Last Theorem without interruption or interference. Fermat too was clearly a loner, separated by geographical distance if nothing else from those who might have been able to understand and appreciate his work. We have spoken of Diophantus and Pythagoras also without any reference to their contemporaries. Were these four men then really lonely geniuses forging new paths alone? Is this how mathematics is done properly, or done best? Let us return to Pythagoras and this time work forward.

The stories about Pythagoras persistently claim that he established or attracted around him a community, or brotherhood, who shared certain religious and philosophical beliefs and perhaps also some mathematical explorations. Unfortunately the stories also claim that the brotherhood was bound to strict secrecy, which of course leaves room for endless speculation about their activities. Even if there is only a grain of truth in such stories, however, it would seem that Pythagoras was charismatic enough to attract followers. Indeed, the fact that his name has survived at all suggests that he was respected and revered in his lifetime, and that he was no hermit.

We are a little better able to place Diophantus, who in Alexandria would have been able to enjoy the company of other scholars. He would also almost certainly have had access to books gathered from other parts of the Mediterranean world in temple or private book collections. It is possible that the problems of the *Arithmetica* were his own invention, but it could equally be the case that he compiled them into a single collection from a variety of other sources, written or oral. One of the recurring motifs of this book will be that mathematics repeatedly passes from one person to another through the spoken word. Diophantus, like any other mathematically creative person, almost certainly discussed his problems and their solutions with a teacher or with students of his own. We should therefore think of him not as a silent figure writing his books in private but as a citizen of a city where learning and intellectual exchange were valued.

Even Fermat, confined to Toulouse and the rigours of full-time political employment, was not quite as isolated as might first appear. One of his friends during his early studies in Bordeaux was Etienne d'Espagnet, whose father had been a friend of the French lawyer and mathematician François Viète. The works of Viète, otherwise rare but thus made available to Fermat, were to have a profound influence upon his development as a mathematician. Another friend, and fellow Counselor in Toulouse, was Pierre de Carcavi, who, when he moved to Paris in 1636, took with him news of Fermat and his discoveries. Through Carcavi, Fermat became known to Marin Mersenne, and through Mersenne he corresponded with Roberval, probably the best mathematician in Paris at the time, and with Descartes in the Netherlands. Later he communicated some of the discoveries arising from his studies of Diophantus to Blaise Pascal in Rouen and to John Wallis in Oxford. Thus even Fermat, far from important centres of learning, was connected into a network of correspondence that stretched across Europe, a virtual community of scholars that later came to be called the Republic of Letters.

When it comes to Wiles, it is much easier to see the cracks in the 'lone genius' story: Wiles was educated at Oxford and Cambridge, and later worked at Harvard, Bonn, Princeton, and Paris, in all of which he was part of flourishing mathematical communities. The mathematical clue that eventually gave direction to his interest in the Last Theorem was picked up from a casual conversation with a fellow mathematician in Princeton; when after five years he needed a fresh breakthrough, he attended an international conference in order to elicit the latest thinking on the subject; when he needed technical help with an important aspect of the proof, he broke his secrecy to a colleague, Nick Katz, and delivered the material in question in a graduate lecture course, though it eventually lost all its listeners except Katz; two weeks before he made the entire proof public in three lectures in Cambridge, England, he asked a colleague, Barry Mazur, to check it; the final proof was checked by six others; and when a flaw was discovered, Wiles invited one of his former students, Richard Taylor, to help him fix it. Further, throughout his years of working on the proof,

Wiles never stopped teaching students or attending departmental seminars. In short, although he spent many hours alone, he was also embedded in a community that allowed him to do so, and which, when required, came to his aid.

Wiles's years of isolation capture the imagination not because they are normal for a working mathematician but because they were exceptional. Mathematics is fundamentally and necessarily a social activity at every level. Every mathematics department in the world contains communal spaces, whether alcoves or common rooms, always equipped with some kind of writing surface, so that mathematicians can put their heads together over the tea and coffee that fuel them. Language or history students rarely write their essays collaboratively, and would not be encouraged to do so, but mathematics students frequently and fruitfully work together, teaching and learning from each other. And despite all the advances of modern technology, mathematics is still primarily learned not so much from books as from other people, through lectures, seminars, and classes.

Stepping stone history

In the outline of the story of Fermat's Last Theorem sketched above, Pythagoras, Diophantus, Fermat, and Wiles appear not only as isolated in their own lives but also from one other, like stepping stones standing out across an otherwise featureless river. If the ivory tower version of history isolates mathematicians from their social groups and communities, the stepping stone version isolates them from their past. Since the past is supposed to be the subject of history, it seems strange to ignore huge chunks of it in this way, but a surprising number of general histories of mathematics are presented in stepping stone style.

Let us then re-examine our story and the gaps in it a little more closely. Just as Pythagoras and Diophantus are somewhat shadowy, so is the space between them. It is possible that

Diophantus had never heard of Pythagoras. He would almost certainly, however, have come across 'Pythagoras' Theorem', not from any writings of Pythagoras, but in the work of Euclid, who lived around 250 BC. Apart from this very approximate date, we know no more about Euclid than about Diophantus a few centuries later, but his master work, the *Elements*, survived to become the longest-running textbook ever, still used in school geometry teaching well into the 20th century. The *Elements* is a comprehensive compilation of the geometry of Euclid's day, with the theorems arranged in careful logical order, and the penultimate theorem of the first book is 'Pythagoras' Theorem', carefully proved by geometric construction. One may reasonably suppose that Diophantus in Alexandria had access to the *Elements*, and it is possible that 'Pythagoras' Theorem' set him thinking about Pythagorean triples. It is equally possible, however, that his inspiration came from other sources that we no longer know about.

The first few centuries between Diophantus and Fermat are almost harder to fill in than those before Diophantus, even in the imagination. We know that Diophantus' *Arithmetica* was originally written in thirteen books, but only the first six survived in Greek; we do not know how or why. (In 1968 an Arabic manuscript was discovered in Iran which claims to be a translation of books IV to VII, but scholars are not agreed as to how accurately the text represents the original.) Fortunately, those six books were preserved for the Greek-speaking world at Byzantium (later Constantinople, now Istanbul), and eventually copies were brought to western Europe. As will be discussed further in Chapter 6, a German scholar known as Regiomontanus saw one of them in Venice in 1462 and believed that it contained the origins of the outlandish subject known to Europeans as algebra. A century later, the Italian engineer and algebraist Rafael Bombelli studied a manuscript of the *Arithmetica* in the Vatican and halted work on his own book on algebra in order to incorporate problems from Diophantus. The first printed edition was published in Basle in 1575, in Latin, translated and edited by Wilhelm Holtzman (Xylander), a humanist scholar, who described the work as 'incomparable, containing the true perfection of arithmetic'. The problems of Diophantus continued to intrigue those who came across them, and in 1621 a new Latin edition of the *Arithmetica* was produced by Claude Gaspard Bachet de Méziriac in Paris. This was the edition that Fermat owned and annotated.

It is not too difficult to fill in the gap between Fermat and Wiles. The Last Theorem, published by Samuel Fermat in 1670, seems not to have attracted any serious attempts in the 17th century, but in the 18th century it came to the attention of Leonhard Euler, the most versatile and prolific mathematician of the period, who made some inroads into the easier cases of it. In 1816, the Paris Academy of Sciences offered a prize for a solution. This inspired the efforts of Sophie Germain, who had some success with certain parts of it and whose work was taken up and extended by others. Beyond that, the problem gradually became widely known and over the years attracted hundreds, if not thousands, of purported solutions, from professionals and amateurs alike. Most of these attempts were both incorrect and useless, but a few led to important mathematical discoveries in their own right, which Wiles would have known about. When he eventually embarked on his own proof, he used some of the deepest mathematics of the 20th century, which was by then known to relate to Fermat's Last Theorem: the Taniyama–Shimura conjecture, made by two Japanese mathematicians in the 1950s, and the Kolyvagin–Flach method, developed by Victor Kolyvagin (Russian) and Matthias Flach (German) in the 1980s. Note again the propensity of mathematicians to write the names of their predecessors into the historical record. Note too the complex web of historical interactions behind a single theorem.

Generally speaking, the further back one goes, the more difficult it is to trace the ground between the stepping stones, not least because much of the evidence has long since been washed away.

But without the attempt, there is no history, only the series of anecdotes on which much of the popular history of mathematics is still too often based.

Elite history

Although we know almost nothing about the lives of Euclid or Diophantus, there are just a few things we can say for certain: that both were well educated and could write fluently in Greek, the intellectual language of the eastern Mediterranean; that both had access to earlier writings on mathematics; that both were able to understand, order, and extend some of the cutting-edge mathematics of their day; and that the mathematics they wrote about had no practical value but was a purely intellectual pursuit. The number of men engaged in such mathematics can never have been great, even in a city like Alexandria. Indeed, it has been estimated that at any one time there were no more than a handful of them anywhere in the Greek-speaking world. In other words, both Euclid and Diophantus belonged to tiny mathematical elites.

A moment's reflection is enough to show how much more mathematics must have been going on than the mathematics they wrote about. Greek society, like every other, had its shopkeepers and housekeepers, farmers and builders, and many others who routinely measured and calculated. We know almost nothing about their methods because such people would have learned and taught mostly by example and word of mouth. Nor were they organized into schools or guilds, though we do know of one named group, the *harpēdonaptai*, or rope-stretchers. By its very nature, their mathematics left few traces. Collections of tokens, or marks scratched in wood, stone, or sand, would have been discarded as soon as they were no longer useful, and were certainly not going to be stored in libraries. In any case, these activities were carried out by people of relatively low social status, and were of little or no interest to the intellectuals of the academies.

When mathematical historians speak of 'Greek mathematics', as they frequently do, they are almost always speaking of the sophisticated written texts that have come down to us from Euclid, Archimedes, Diophantus, and others, not of the common or garden mathematics of the *hoi polloi*. Recently this has begun to change. Historians have started to acknowledge that elite Greek mathematics had its roots in the practical and everyday mathematics of the eastern Mediterranean, even if later writers distanced themselves from those roots by developing a more formal and 'useless' kind of mathematics.

There is something else to be wary of in the catch-all phrase 'Greek mathematics'. Diophantus lived in Alexandria in Egypt; Archimedes lived in Syracuse, on the island of Sicily; Apollonius, another of the great 'Greek' mathematicians, lived in Perga, in the region that is now Turkey; in other words, although all wrote in Greek, none of them came from the area that we now know as Greece. Indeed, Diophantus, for all we know, could have been African born and bred. Nevertheless, 'Greek mathematics', so highly revered by Renaissance Europeans, has come to be thought of as essentially 'European'. The absurdity of incorporating Alexandria into Europe becomes all the more apparent when we think of the exclusion of Spain, at the opposite end of the continent. Spain came under Islamic rule early in the 8th century and consequently enjoyed the rich culture and learning of the Islamic world. Yet one frequently reads that Arabic numerals were introduced to Europe by Fibonacci, writing in Pisa in Italy in the early 13th century, as though their use in Spain for two centuries before that counted for nothing, and as though Spain were somehow not a proper part of Europe. Those promoting the cause of elite mathematics have naturally tended to assimilate into their histories whatever would give their subject authority and respectability, regardless of other inconvenient facts.

Wherever mathematics is practised, we are likely to find a few advanced and highly respected practitioners but many more whose names will never enter any history book. If we re-examine the situation in Fermat's day, we will find it hardly any different. During his lifetime, France was exceptionally rich in elite mathematical activity: one can think of as many as three or four Parisians who could have kept up with Fermat. At a generous estimate, there were perhaps as many again in the Netherlands and Italy together, and even one or two in England, but no more than that. Yet mathematical activity lower down the social scale was more widespread than one might expect. Recent electronic searches of digitized material have shown that as many as a quarter of the books published in England in the 16th and 17th centuries mentioned mathematics in one way or another, if only in passing. Further, there was a steady increase in books aimed at tradesmen or craftsmen who wanted to acquire basic mathematical skills.

Before ending this chapter, let us look in a little more detail at one of them: there is after all no better way of exploring the history of mathematics than to delve into the original sources. Robert Recorde's *The Pathway to Knowledg* was published in England in 1551, about 50 years before Fermat was born. For much of his life, Recorde practised as a physician. In 1549 he was appointed controller of the Bristol mint, and two years later surveyor of silver mines in Ireland. Unfortunately he made political enemies in this period and ended up in the King's Bench prison in London, where he died in 1558 at the age of 48. It was also during this time, however, that he published most of the mathematical works for which he is now remembered. Educated at Oxford and Cambridge, Recorde was fluent in Latin and Greek but made the bold choice of writing his mathematical texts in English. In particular, he aimed to make the mathematics of Euclid, one of the most elite of mathematicians, available to the common man. This was no easy task: for one thing, most English workmen, though they may have been adept enough with plumb lines and rulers, had never heard of a formal subject called 'geometry'; for another, there were simply no words in English for technicalities like ‘parallelogram’ or ‘sector’. Recorde addressed both problems with imagination and skill.

In a lengthy preface he described the classes of men for whom geometry was ‘much necessary’, from those of humblest social status upwards. At the bottom were the ‘unlearned sort’ who worked the land. Even these men, Recorde argued, had an instinctive grasp of geometry, otherwise their ditches would collapse and their haystacks would topple. Moving upwards to tradesmen, Recorde supplied a long list, in verse, of those to whom geometry was indispensable: merchants, navigators, carpenters, carvers, joiners, masons, painters, tailors, shoemakers, weavers, and more, concluding

> That never was arte so wonderfull witty
> So needefull to man, as is good Geometry.

Recorde also deemed geometry to be indispensable in the professions of medicine, divinity, and law, though his arguments became rather more artificial and less convincing as he climbed the social scale.

Recorde’s empathy with the common man is clearest when he comes to the geometry itself: his exposition is a model of good pedagogy, expressed in plain language with lots of examples and helpful diagrams. Quite early on he teaches Euclid’s ruler and compass construction of a right angle. In case this should prove too difficult, however, he has an alternative suggestion: take a line and mark off three, four, and five units respectively, and then use those lengths to create a triangle. The angle between the short sides will be a right angle. This is no classical Euclidean construction: it is a method for practical men, for rope-stretchers.

For the 21st century, we could make a far longer list than Recorde could of those who use mathematics in their everyday life, in

1. 'Colourwash' by Tatjana Tekkel Peppé who, by her own account, is no good at mathematics

school, the home, or the workplace. I am thinking of my mother Irene, who at the age of 89 trusted neither banks nor computers, but tallied every penny of her household expenditure in carefully ruled notebooks; or of my friend Tatjana, who repeatedly tells me she was no good at mathematics at school but who creates intricately designed quilts (see Figure 1). She can certainly handle right-angled triangles. Indeed, her instinct for tessellation and proportion qualify her, perhaps, as a modern-day representative of the *harpēdonaptai*.

Elite history does not have any space for Irene or Tatjana: women, in particular, have to rise at least to the level of Sophie Germain before they are taken seriously. Yet without people who do and teach mathematics at every level, the elite could not flourish. Behind the outposts occupied by Wiles, Fermat, or Diophantus, there stretch vast hinterlands of mathematical activity that have been all too little explored in general histories of the subject. Part of the purpose of this book is to redress the balance and to reclaim mathematics for the man, woman, and child in the street, to revisit the history of mathematics from some new perspectives.

Chapter 2
What is mathematics and who is a mathematician?

In the previous chapter, I assumed that readers would take 'mathematics' to be more or less the subject they studied under that name at school, and 'mathematicians' to be those people who continue with it into adult life. History, however, requires us to think about both terms more carefully. So does experience: when as a school teacher I found myself in a single morning delivering lessons on percentages, circle theorems, and differential calculus, I was forced to ask myself how this unlikely collection of topics had come together under the single heading of 'mathematics'. Most people would probably agree with the rather general statement that mathematics is based on properties of space and number, but how then would they regard the popular puzzle Sudoku? Is it a mathematical pursuit or not? I have heard expert mathematicians argue that it is or that it isn't, with equal vehemence either way.

Let us go back to a beginning. The Greek word *mathemata* simply meant 'what has been learned', sometimes in a general way, at other times connected more specifically to astronomy, arithmetic, or music. The Greek word entered into the etymology of the modern word 'mathematics' and its cognates in other European languages (*mathématiques*, *Mathematik*, *matematica*, or, in US English, *math*). The meanings of the word 'mathematics', however, slipped and twisted through many variations over the centuries, as we shall see shortly. And that is looking at the matter only from a European perspective. If we go back one or two thousand years, before European culture became dominant, can we find words equivalent to our 'mathematics' in Chinese, Tamil, Mayan, or Arabic? If so, what writings and activities did those words cover? To investigate that question thoroughly would be a lifetime's work for an army of scholars, but here, as elsewhere in this book, some case studies will serve to illustrate the questions that need to be asked and the kind of answers that can arise.

Tracing some meanings of *suàn*

From histories composed by Chinese government officials for the period from shortly before 200 BC to AD 200 (the Qín and Hàn periods) it is possible to discover the names of just over 20 people who were said to have been skilled in some aspect of *suàn* 算. As a noun, *suàn* can mean a set of short rods, made of wood, metal, or ivory, which are manipulated on a flat surface to record the numbers in a calculation; it can also mean the act of using the rods. Here then is evidence of mathematical activity, but we still do not know very much unless we can discover what kind of calculations were carried out.

For many of the practitioners named in the official records, it appears that *suàn* was closely associated with the astronomical or calendrical systems known as *lì* 曆. All pre-modern societies used the positions of the Sun, Moon, and planets to determine appropriate times and dates for religious rituals or the planting of crops, so those who could make correct predictions from astronomical data were indispensable to rulers and governments. There are thus repeated associations of *suàn* with *lì* in the histories of early imperial China. The same records also show, however, that *suàn* was relevant to more earthly matters, the calculation of profit, and the distribution of resources.

In the early 1980s, a new source was discovered for the period around 200 BC, one that sheds further light on the use of *suàn* at that time. The text known as the *Suàn shù shū* 算數書 is inscribed on 190 bamboo strips, each about 30cm long, which were originally joined side by side with knotted string so that they could be rolled together like a mat. The final word *shū* means 'writings' or sometimes 'book'. The middle word, *shù*, may be broadly interpreted as 'number'. What is most relevant for our purposes, however, is the meaning of the combination *suàn shù*. The writings on *suàn shù* contain around 70 problems with instructions for solving them. These include: multiplying whole numbers and fractions; sharing out profit according to the amounts put in by different contributors; allowing for waste in the production of commodities; calculating total cost from the price of a given amount; calculations of tax; finding amounts of ingredients in a mixture; converting an amount of raw material to a number of finished products; checking times taken for a journey; calculation of volumes and areas; conversion of units.

Thus for the most part the problems of the *Suàn shù shū* are based on everyday activities and transactions. It is written in a very direct style: for each problem the writer poses 'question', 'result', and 'method'. Here as examples are two 'customs-post problems' from the second chapter:

> A fox, a wild-cat and a dog go through a customs-post; they are taxed 111 coins. The dog says to the wild-cat, and the wild-cat says to the fox 'Your skin is worth twice mine; you should pay twice as much tax!' Question: how much is paid out in each case: Result: the dog pays out 15 coins and $\frac{6}{7}$ coins; the wild-cat pays out 31 coins and 5 parts; the fox pays out 63 coins and 3 parts. Method: let them be double one another, and combine them [into] 7 to make the divisor; multiply each by the tax to make the dividends; obtain one for [each time] the dividend accommodates the divisor.

And perhaps more realistically:

> A man is carrying hulled grain – we do not know how much – as he passes through three customs posts. [Each] post takes a duty of 1 in 3. After leaving he has one *dǒu* of hulled grain left. Question: when he started going, how much hulled grain did he bring? Result: The hulled grain he brought was 3 *dǒu* 3 *shēng* and $\frac{3}{4}$. Method: Set out one, and thrice double it to make the divisor. Again set out one *dǒu* of hulled grain and 3-fold it. Again three-fold it and [multiply by] the number of passes to make the dividend for it.

The answers are correct but the descriptions of the 'method' are not very enlightening, and it is likely that they were meant to be supplemented by oral explanation. The instructions are given only for the particular numbers in the stated question, but a trained reader would be able to adapt them to any similar problem, so in that sense they provide a general technique. There is no expectation in the text, however, that the reader should understand the reasoning behind the method, only that he should be able to apply it.

Similar problems and others appear in a later text, the *Jiǔ zhāng suàn shù* 九章算術, writings on *suàn shù* in nine sections, commonly known in English as the 'Nine Chapters'. The official histories show that the text was in use by the beginning of the 2nd century AD. As with Euclid's *Elements* from some three or four centuries earlier, however, we have no information about the author or composition of the 'Nine Chapters', nor the original text. The only version that has come down to us is the one given by Liú Huī 劉徽 in AD 263. Until the transcription and publication of the contents of the *Suàn shù shū* in 2000, the 'Nine Chapters' was the earliest extensive text devoted to *suàn*. The discovery of the *Suàn shù shū* therefore not only makes possible important textual comparisons, but also offers historians a much deeper knowledge of the uses of *suàn* in the early years of imperial China.

It is clear even from this very short account that the word *suàn* was not associated with any overarching subject that we can capture with the single word 'mathematics'. Instead it denotes techniques and skills that could be put to use in a range of contexts, from applications to *lì*, the astronomical reckonings required at court, to the more mundane *suàn shù*. Turning now to the Latin West, can we find a similar range of practices associated with the word 'mathematics'?

Tracing some meanings of 'mathematics'

Around AD 100 the Roman writer Nicomachus listed four disciplines concerned with multitude and magnitude: arithmetic, music, geometry, and astronomy. For Nicomachus, arithmetic, the study of multitudes (or numbers), and geometry, the study of magnitudes, were the most fundamental; music was the science of multitudes in relation to one another, while astronomy dealt with magnitudes in motion. Four centuries later, the philosopher Boethius described these disciplines collectively as the *quadrivium*. Together with the *trivium* of grammar, logic, and rhetoric, they made up the seven liberal arts of the medieval academic curriculum. Boethius himself wrote treatises on arithmetic and music that were studied in European universities throughout the medieval period. Some writings on geometry were also ascribed to him, but their true authorship is uncertain: Boethius, like Pythagoras, became something of a mythical figure to whom later work could usefully be attributed.

Arithmetic and geometry remain at the heart of mathematics (they are the activities, we may recall, practised by Irene and Tatjana) but astronomy and music have now gone their separate ways. The break came in the 17th century when it became increasingly difficult to reconcile mathematical theory with musical practice, and when astronomy struggled to free itself of its long associations with astrology to become a respectable subject in its own right.

In any case, by the time of the Renaissance the four-fold division of Nicomachus was too constrained to accommodate the many new kinds of mathematical activity that were beginning to emerge in response to the rapid growth of wealth, trade, and travel. John Dee, in a preface to the first English translation of Euclid's *Elements*, in 1570, set out a 'groundplat', or plan, of the mathematical arts and sciences (see Figure 2). Arithmetic and geometry remain the key components, but by now geometry, which answers the questions 'How farre?', 'How high or deepe?', 'How broad?', has given rise to 'geographie', 'chorographie', 'hydrographie', and something called 'stratarithmetrie'. Further, there is a long list of subjects regarded as 'derivatives' of both arithmetic and geometry, including 'astronomie' and 'musike' among many others. The modern reader will have some idea of what was meant by 'perspective', 'cosmographie', 'astrologie', 'statike', 'architecture', and 'navigation' but will probably be as bemused as contemporary readers might have been by 'anthropographie', 'pneumatithmie', 'archemastrie', and several other uncommon branches of learning. Indeed, the obscurity of the subject matter and the neat divisions under subheadings and sub-subheadings suggest that Dee's systematization, like the much simpler scheme of Nicomachus or Boethius, was a philosophical exercise rather than a genuine classification of existing practices.

How then are we to find out more precisely what mathematical activity in western Europe consisted of during the centuries between AD 500 and 1500? Can we carry out the same kind of study for 'mathematics' as we did for *suàn*, discovering the meanings of the word by examining the contexts in which it was used? There are many more surviving texts from western Europe over this period than there are for early imperial China, so a full survey is impossible, but as a first approach we will examine a mathematical history compiled by the Dutch scholar Johann Gerard Vossius, his *De scientiis mathematicis*, published in Amsterdam in 1649, in particular as it relates to British writers.

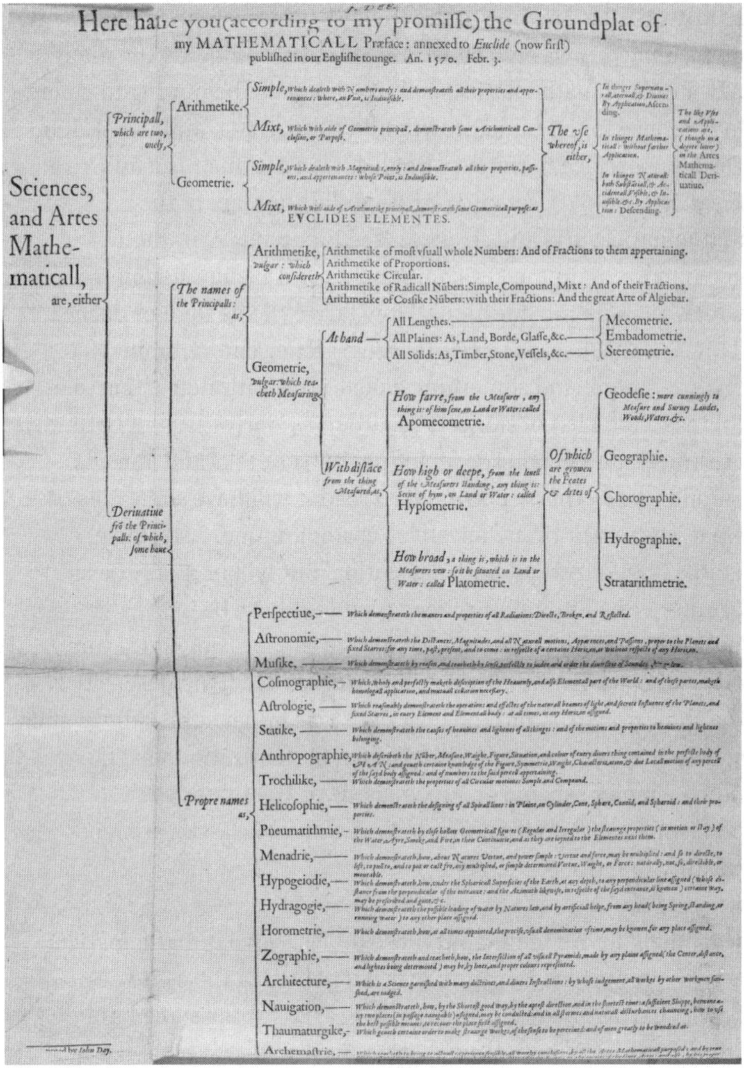

2. John Dee's 'groundplat' from his preface to Euclid's *Elements*, 1570

It may seem strange to turn to a Dutch scholar for information on British intellectual history, but much of Vossius's account of British authors was based on earlier work by the English antiquary John Leland. In 1533, shortly before the dissolution of the monasteries, Leland was commissioned by Henry VIII to search the libraries and colleges of the realm and to list their collections. Over the next two or three years, he listed the holdings of some 140 religious foundations. The subsequent dispersal and loss of books grieved him greatly: in 1536 he complained to Thomas Cromwell that 'the Germanes perceiving our desidiousness [indolence] and negligence, do send dayly young Scholars hither, that spoileth them, and cutteth them out of Libraries'. Leland provided the last and most comprehensive record of what the libraries had contained. He intended to compile a dictionary of British writers, containing some 600 entries, but sadly became insane before it was quite completed. His invaluable work was recognized by other historians, however, and a great number of later writers, including Vossius, drew directly or indirectly on his findings.

The earliest English writer mentioned by Vossius was Bede, writing about AD 730, who was listed under both 'astronomy' and 'arithmetic'. Bede, who spent most of his life in the monastery at Jarrow in the north-east of England, is well known as a biblical commentator and church historian, but few would now count him an astronomer. The writings ascribed to him, however, are described as being on the Moon and its cycles, the date of Easter, the planets and zodiac, the use of the astrolabe, and calculation of the vernal equinox. Some of these writings may have been mistakenly ascribed to Bede by later commentators, but he was certainly much concerned with the date of Easter, which was as crucial to Christians as the correct timing of the winter solstice had been to the early Chinese emperors. It was not an easy calculation either: Easter had to fall on the first Sunday following a full Moon after the spring equinox, and so correct calculation of the date required an understanding both of lunar and solar cycles, which are not naturally correlated. The presence of two Christian traditions in northern England, Irish and Roman, had led to conflicting dates, a situation that had eventually been resolved at the Synod of Whitby in 664. Bede may not have carried out the necessary calculations himself, but he knew what was at stake.

Calculations concerned with ecclesiastical time-keeping eventually became known by the name of *computus*, and remained essential throughout the medieval period.

After Bede and his follower Alcuin, no further English names appear in Vossius's account for more than four centuries, until we meet Adelard of Bath around 1130. Adelard, who appears to have travelled in France, Sicily, and Syria, was one of the earliest translators of parts of Euclid's *Elements* from Arabic to Latin, and was also said to have written on the astrolabe.

Only for the 13th and 14th centuries do the names (and their supposed dates) begin to appear with increasing frequency, all of them under 'astronomy' or 'astrology': John Sacrobosco (1230), whose writings on the Earth and its place in the universe remained a key part of the university curriculum for four centuries; Roger Bacon (1255), described as an astrologer; Walter Oddington (1280), said to have written on the motion of the planets; Robert Holcot (1340) of Northampton, said to have written on the motion of the stars; John Eastwood (1347), astrologer; Nicholas Lynne (1355), astrologer; John Killingworth (1360), astronomer; Simon Bredon (1386), said to have written on medicine, astrology, and astronomy; John Summer (1390), astrologer, and so on. Then in the 15th century the names begin to fade out again. Clearly the 14th century was a high point of astronomical and astrological studies, one contributory factor perhaps being the terrible shock of the Black Death in 1348. Many of those mentioned belonged to religious orders, Franciscan, Dominican, or Carmelite. Many were also linked to Oxford, particularly to Merton College, and some of their writings are safeguarded to this day in Oxford libraries. All of them crossed and re-crossed the fluid boundaries between astronomy and astrology.

In contrast to this galaxy of astronomers, no English writers appear in any of Vossius's chapters on music, optics, geodesics, cosmography, chronology, or mechanics, and only Gervase of

Tilbury and Roger Bacon are mentioned under geography, as map-makers. Thus, looking back from a 16th-century standpoint to the mathematical writings of medieval England, the dominant themes are *computus* and astrology.

For other regions of Europe, however, the picture would have been different. In Italy, for instance, situated in the heart of the western Mediterranean, trade was more extensive and more complex than in northern Europe, and the 13th century saw the establishment of *abacus* schools to train boys in commercial arithmetic and even a little rudimentary algebra (solving some basic equations). The seminal text was the *Liber abaci* (*Book of Abacus*) of Leonardo of Pisa, later known also as Fibonacci. The *Liber abaci* contains hundreds of commercial problems. Here are two of them:

> Four men made a company in which the first man put $\frac{1}{3}$ of a whole, another put $\frac{1}{4}$, a third put $\frac{1}{5}$, and a fourth truly put $\frac{1}{6}$, and they had together a profit of 60 soldi; it is sought how much each held of it. The problem truly is the same as was said about four men who buy a pig for 60 soldi of which the first wishes to have one third of the pig, the second one fourth, the third one fifth, and the fourth one sixth....

Leonardo himself has pointed to two versions of this question; it is also mathematically equivalent to the fox, dog, and wild-cat problem from the *Suàn shù shū*. The next problem reflects the concerns of contemporary Italy and is typical of hundreds of questions on conversion of currencies or materials. At the same time, it shows that some ten centuries after Diophantus, arithmetic of another kind was still thriving in Alexandria.

> Also 11 Genoese rolls [of cloth] are worth 17 carats in Alexandria; how much are 9 Florentine rolls worth? Because the 11 rolls and the 9 rolls are not the same units of weight, you make Florentine rolls of the 11 Genoese rolls, or you make Genoese rolls of the 9 Florentine rolls so that both will be either Florentine rolls or Genoese rolls; but

> because you can easily make Florentine rolls, each Genoese roll is $\frac{1}{6}2$ Florentine rolls, you will multiply the Genoese rolls by $\frac{1}{6}2$ to make $\frac{5}{6}23$ Florentine rolls...

For all their learning, Vossius and his sources in northern Europe had never seen the *Liber abaci*; Vossius knew of it only by hearsay and got its date wrong by two centuries. Mathematical activity could be very localized.

It was also time-related. For the medieval period, most of the headings later invented by Dee and Vossius would have been largely redundant, at least for England. By the end of the 16th century, as Britain too entered into the wider world, that was no longer the case. Thomas Harriot, working around 1600, left writings on optics, ballistics, alchemy, algebra, geometry, navigation, and astronomy. Meanwhile, his contemporary Simon Stevin in the Netherlands published on a similar range of subjects, but with navigation replaced by the more pertinent problems (to him) of locks and sluices. *Computus* and astrology had given way to the mathematical activities of a new world order.

What is mathematics?

What then has mathematics been historically, if indeed there has ever been such an entity? It should be clear by now that mathematical activity has taken many forms, only loosely connected by the fact that they require some kind of measurement or calculation. A more precise answer must be heavily dependent on time and place. There are a few common threads: all organized societies need to regulate trade and time-keeping, which were very roughly speaking the aims of *suàn shù* and *suàn lì*, respectively, in early imperial China, or of *abacus* and *computus* in 13th-century Europe. The practitioners of these various techniques, however, were likely to have been of very different social status. *Suàn shù* and *abacus* teachings were intended for merchants or officials, whereas *suàn lì* or *computus* were the province of high-ranking specialists in China, and of monks and scholars in medieval Europe. A separation of status and respect between those sufficiently educated to engage in 'higher' mathematics, which usually requires a certain level of abstract thinking, and the tradesmen or craftsmen who work with 'common' or 'vulgar' mathematics, has recurred in different contexts over many centuries.

As societies become more complex so too do their mathematical requirements. The long list of headings proposed by Dee, even if some were redundant, indicates a wide range of activities in which mathematical expertise was invoked. These subjects were collectively known as 'mixed mathematics', suggesting that mathematics was an integral part of each of them (not quite the same as the later concept of 'applied mathematics', where mathematics is used to analyse subjects outside itself).

There is no reason to suppose that the lessons learned from early imperial China and medieval Europe do not extend to other societies too: that there is no single body of knowledge that we can conveniently call 'mathematics' but that we can identify many mathematical disciplines and activities. And which particular ones are considered most relevant or prestigious has always been a matter of time and place.

Who is a mathematician?

Now that we have begun to identify the range of activities that have constituted mathematics, can we say who does or does not count as a mathematician? All four of Pythagoras, Diophantus, Fermat, and Wiles are commonly described as mathematicians, and the first three, being dead, have made it into a standard reference work, the *Biographical Dictionary of Mathematicians*. None of them, however, would have recognized the label they have been given. We have no idea how, if at all, Pythagoras would have described himself. Diophantus would probably have thought of himself as an arithmetician, not as a practitioner of everyday arithmetic of the *suàn shù* or *abacus* kind, but of the 'higher arithmetic' that investigates some of the more obscure and difficult properties of the natural numbers. Fermat, on the other hand, would have called himself a *géomètre*, geometry by then being the most authoritative and respectable branch of the quadrivium. This remained a standard description of an academic mathematician in France well into the 19th century. Of the four, only Wiles, I suggest, would unreservedly call himself a mathematician.

Today the discipline of mathematics is highly respected, even revered, but from what has been said already in this chapter it can easily be seen why this has not always been the case. John of Salisbury in the 12th century claimed that the practice of *mathematica*, the foretelling of the future from the positions of the stars and planets, arose from a fateful familiarity between men and demons, and along with chiromancy (palm-reading) and augury (interpreting the flight of birds) was a source of evil. In 1570, Girolamo Cardano, medical practitioner and author of one of the leading algebra texts of the Renaissance, was imprisoned for casting a horoscope of Christ; Thomas Harriot, arrested in 1605 on charges of association with perpetrators of the Gunpowder Plot, was questioned not so much about the plot itself but about the fact that he had a horoscope of James I pinned to his wall; and late in the 17th century John Aubrey wrote of the country clergyman and mathematics teacher William Oughtred that 'The country people did believe that he could conjure'. In pre-modern Europe, the practice of 'mathematics' was not without its dangers, to the practitioner as much as to his supposed subjects.

In fact, the word 'mathematician' began to be used regularly in English mathematical writings only from 1570 onwards. At first, it was used mainly for foreign authors, but later in two curiously unrelated contexts, for gunners or astrologers. After the Restoration in 1660 it came to be used more generally for writers on arithmetic or geometry but also still for astrologers; at the same time, the predictions of 'the mathematicks' became a regular subject of satire and ridicule. The longstanding and persistent association of mathematics with astrology helps to explain why academics preferred to avoid the term. When Henry Savile founded two mathematical chairs in Oxford in 1619, they were in Geometry and Astronomy, respectively, with strict directions that the latter should not include judicial astrology. To this day Cambridge hosts a Lucasian Professor of Mathematics, but Oxford's equivalent is the Savilian Professor of Geometry. And unless it should be thought that the association of mathematics with prediction and influence was only a European phenomenon, it is worth bearing in mind that the modern Chinese term for mathematics, *shù xué* 數學, has traditionally meant the study of numbers in the context of divination.

In short, 'mathematicians', as we now understand the term, are a modern European invention. In the long history of mathematical activity, they have existed for little more than a blink of the eye, and if we are to appreciate mathematical history properly it is crucial not to project their image back onto the past. For that reason, historians prefer to use more precise descriptions like 'scribe', 'cosmographer', or 'algebraist', or more general terms like 'mathematical practitioner'. One thing is certain: the history of mathematics is not the history of mathematicians.

Chapter 3
How are mathematical ideas disseminated?

The previous chapter carried out some broad surveys of mathematical activity at different times and places. This is one way of studying the history of mathematics: determining what people actually did. The historian always wants to ask further questions, however: not just what people knew but how they communicated it to one another and to those who lived after them. How are mathematical ideas passed from one person to another, from one culture to another, or from one generation to another? (Recall the questions first raised in Chapter 1: how did Fermat know about Diophantus, or Wiles about Fermat?)

An extension of these questions is to ask how historians themselves can know about the mathematics of the past: what sources do we have, how have they come down to us, how reliable are they, and how can we learn to read them? This chapter will examine the way mathematical ideas have sometimes traversed long distances of time and space, but also how, often, they have not.

Fragility, scarcity, and obscurity

Those who have comfortably assumed that mathematics began with Pythagoras may now suffer a slight sense of vertigo on discovering that sophisticated mathematics was already being practised more than a thousand years earlier in Egypt and in the region that is modern Iraq. The Egyptian and Babylonian civilizations of the 2nd and 1st millennia BC existed in relative proximity to each other, but we know very much more about the mathematics of the latter than the former for the very simple reason that the clay tablets used as writing material along the Tigris and Euphrates were robust and durable whereas the papyri of the Nile region were not. Thousands of tablets have been excavated from Iraq, many with mathematical content, and thousands more probably remain buried if they have not been crushed by tank treads or looted in the chaotic aftermath of recent wars. For Egypt, on the other hand, the number of surviving mathematical texts and fragments can be counted on the fingers of three hands, and those are scattered across a thousand years of history. The equivalent for Britain would be a few texts from around the time of the Norman conquest and a few more from the 19th century. Clearly the surviving Egyptian texts provide only the most meagre insight but at the same time leave ample room for speculation and fantasy about Egyptian mathematical activity.

For India, south-east Asia, and South America, the situation has been much the same as for Egypt: the climate has rapidly destroyed natural materials like wood, skin, or bone, so that historians have had to do the best they can with very few texts, poorly preserved. Clearly the paucity of material distorts our picture of the past. We must ask whether what survives is typical of what has been lost, knowing that a single new discovery (like the *Suàn shù shū* in China) could radically alter our perceptions of an entire mathematical culture. At the same time, the lack of texts has perhaps had some benefits in that it has forced historians to broaden their search for sources. Administrative records, for instance, can reveal the counting and measuring that were carried out in everyday life. Archaeological evidence has improved our knowledge of how buildings were planned and constructed and what calculations must therefore have gone into them (for we have no direct evidence of any calculations that went into the building of Stonehenge or the Pyramids). Sources as varied as pictures, stories, or poems may also include hints of contemporary mathematical knowledge.

Many ancient texts were written in scripts and languages that are now extinct, and the process of translating them is fraught with difficulties. The number of scholars with the requisite language skills who are also brave enough to engage with mathematical material remains very small indeed, and their task is exceedingly delicate. Any translation from one language to another risks destroying something of the essence of the original, but mathematical translation introduces a further difficulty: how to make the technical concepts of another culture comprehensible to a modern audience. What can the ordinary reader make, for example, of the following passage from the Indian *Brāhmasphuṭasiddhānta* from AD 628:

> The height of a mountain multiplied by a given multiplier is the distance to a city; it is not erased. When it is divided by the multiplier increased by two it is the leap of one of the two who make the same journey.

To understand this problem, the reader needs to know that one traveller descends a mountain and walks along the plain to a city, while the other magically leaps from the mountain top to a greater vertical height and flies along the hypotenuse, but in doing so covers just the same distance. For a student at the time, this problem may have been one of a standard type (another version of it has monkeys leaping up trees) and was probably elucidated through oral explanation, but for a 21st-century reader with no knowledge of Sanskrit or 7th-century Indian mathematical conventions, it is at first sight simply baffling.

Thus a literal translation of a raw text is not likely to convey very much to a non-specialist. An age-old way of getting round this problem is for translators (or copyists) to add annotations or explanatory diagrams: all important mathematical texts have accrued layers of commentary in this way. Another method is to translate the text into modern mathematical notation. The reader who wishes to try this for the problem of the two mountain travellers will probably find that it makes it much clearer. The use of modern algebraic notation can be helpful as a preliminary way in to understanding the mathematics of the past but should never be mistaken for what the original writer was 'really' trying to do, or what he would have done with the advantage of a good modern education. At best, such modernization obscures the original method and at worst can lead to serious misunderstandings.

The surviving Egyptian texts of the 2nd millennium BC, for example, are written in hieratic, a cursive script that replaced hieroglyphics in everyday use from about 2000 BC onwards. They were translated into English or German in the early 20th century and for many years those translations remained standard. Unfortunately, however, the contents were translated not only into modern languages but also into modern mathematics. It is often stated, for instance, that the Egyptians used a value of 3.16 for the number we now denote by π, the multiplying factor that gives the area of a circle from its radius-squared (as a modern formula, we may write $A = \pi r^2$). When we examine the texts on which this claim is based, we find that they do not expect the reader to multiply the radius-squared by any number at all. Instead they instruct him to find the area by reducing the diameter by $\frac{1}{9}$ and then squaring it. A bit of pencil and paper calculation shows that this gives the area of the circle as $\frac{256}{81}$ times the radius-squared, hence the magic value of $\frac{256}{81} = 3.16\ldots$ But 'reducing and squaring' is not the same as 'squaring and multiplying', even if it gives very nearly the same answer: the process is quite different, and processes are precisely what historians need to be concerned with if they are to understand the mathematical thinking of earlier cultures.

The story of translation of Babylonian texts has been similar. Here the languages are Sumerian, unrelated to any surviving language, and Akkadian, a precursor of Arabic and Hebrew; and the writing is cuneiform, imprinted into wet clay with a sharpened reed. A large number of mathematical texts were translated and published during the 1930s by Otto Neugebauer and François Thureau-Dangin, and thereafter for many years the job was thought to be more or less done. These early translations, however, all too often turned Mesopotamian calculation techniques into their modern algebraic equivalents, obscuring the true nature of what the original scribe was actually thinking and doing, while at the same time making the calculations look rather primitive. Only since the 1990s have many of the tablets been translated afresh, with much greater care for the original language. Words that mean literally 'break in half' or 'append', for instance, convey physical actions that are quite lost in the abstract translations 'divide by 2' or 'add', and give us a much better insight into the way problems were understood or taught.

Reading and translating texts is only one part of the work of historians of ancient mathematics, albeit an important one. The other is to interpret them within their own context. Sometimes this is simply impossible: many Middle Eastern texts excavated or rediscovered in the 19th century, including almost all the extant Egyptian hieratic texts and hundreds of Old Babylonian cuneiform tablets, changed hands in the antiquities market carrying no known provenance. Unfortunately, many looted or stolen objects are still bought and sold this way today.

The fragility and scarcity of mathematical texts improves only a little as we move forward from the ancient world to the medieval period. Even documents deliberately preserved in libraries are not always secure. There are varying accounts, now impossible to confirm, of the destruction of the library at Alexandria in times of conflict, and certainly it would have been as vulnerable to fire as any pre-modern library housing books or manuscripts. Readers at

Oxford's Bodleian Library are still required to swear an oath promising 'not to bring into the Library, or kindle therein, any fire or flame, and not to smoke in the Library', a reminder of the days when such activities could prove as lethal to books as to people.

We have already seen the efforts of John Leland to record the contents of monastic libraries, but he could not preserve more than a fraction of the collections themselves when those libraries were eventually destroyed and their holdings dispersed. There were other dangers too: Merton College in Oxford threw out a great many manuscript books during the 16th century as it modernized to printed texts, and though some were rescued by alert collectors, there must have been many that were not. And John Wallis in 1685, like Leland more than a century earlier, complained bitterly about the theft of valuable material: two 12th-century prefaces, he wrote, had 'lately (by some unknown hand) been cut out, and carried away' from a manuscript in Corpus Christi College. He hoped that 'Who ever hath them, would do a kindness (by some way or other) to restore them', but he hoped in vain: the prefaces are still missing.

Private collections of papers were also vulnerable: John Pell, worrying in 1644 about the mathematical papers of his recently deceased friend Walter Warner, wrote:

> I am not a little afraid that all Mr Warner's papers, and no small share of my labour therein, are seazed upon, and most unmathematically divided between the sequestrators and creditors, who will, no doubt, determine once in their lives to become figure-casters, and so vote them all to be throwen into the fire.

Printed books are just as susceptible as manuscripts to fire, flood, insects, and human carelessness, but because more copies are produced more are likely to survive. Those that come down to us, however, are unlikely to be typical of what once existed. An expensive volume from a gentleman's library is more likely to last than a tradesman's well-thumbed ready-reckoner, but probably tells us less about what was actually read and used.

Constructing a true understanding of the past often feels like trying to put together a jigsaw puzzle in which most of the pieces are missing and there is no picture on the box. Nevertheless, remarkably, we do have mathematical texts that have survived for centuries, even millennia. For the most part, their contents are of purely historical interest: no-one now calculates with Egyptian fractions, except as a school exercise, and the only vestiges of the Babylonian sexagesimal system are in our otherwise curious division of an hour into 60 minutes and a circle into 360 degrees. Other texts, however, have remained very much alive through continued use and translation, and occasionally it is even possible to trace an almost continuous line of descent from past to present. The outstanding example must be Euclid's *Elements*, which has already been mentioned more than once, and without which no history of mathematics can be complete. A study of what is sometimes called the 'transmission history' of the *Elements* tells us much about how mathematical ideas from the past may be preserved, amended, and passed on.

Preservation through time

The remarks made above on the fragility of Egyptian sources apply just as much to texts from the ancient Greek-speaking world, which were also written on papyrus. We assume, from contemporary references to some of his other works, that Euclid wrote around 250 BC. Yet the earliest surviving text of the *Elements* is from AD 888. That represents over a thousand years of copying and re-copying, with all the scope that entails for error, change, and 'improvement'. How can we know that the text we have now is in any way faithful to the original? The answer is that we can not. In the case of the *Elements*, we have extensive commentaries from later Greek writers, Pappus (AD 320), Theon

(AD 380), and Proclus (AD 450), which tell us how the text appeared in the 4th or 5th century BC. These men lived much closer to Euclid's time than we do, but still several centuries after the *Elements* was first written. Otherwise, the only way historians can approach the original is to construct a 'family tree' of surviving manuscripts, by observing, for example, where mistakes or alterations have been copied from one text to another. In this way they would hope to arrive back at a 'master copy', but it is painstaking work with no guarantee that it will take one back to a true and unique source.

That earliest surviving manuscript of the *Elements*, from AD 888, is written in Greek and was preserved in Byzantium. But as Islam spread into the old Greek-speaking regions of the Mediterranean, the text was also translated into Arabic. One can imagine what difficulties the early Islamic translators might have encountered by comparing their task with that of Robert Recorde several centuries later: it is unlikely that Arabic, the language of nomadic tribes, contained ready-made words for the abstract concepts of Euclidean geometry. Nevertheless, Arabic translators saved many texts from extinction.

Most medieval translations of the *Elements* into Latin were thus made not from Greek, a language that had by then all but died out in western Europe, but from Arabic sources in Spain or Sicily. Adelard of Bath, whom we met in the previous chapter, was one such translator, and there were several others in the 12th century, scholars from northern Europe who travelled south in search of the learning that could be found there. Eventually, as knowledge of Greek was slowly revived, translations were also made directly from Greek sources.

Once printing was established, in the 15th century, Euclid's *Elements* was finally secured for posterity. It was amongst the first mathematical books to be printed, in a beautiful edition of 1482 that continued the traditions of manuscript production: there is

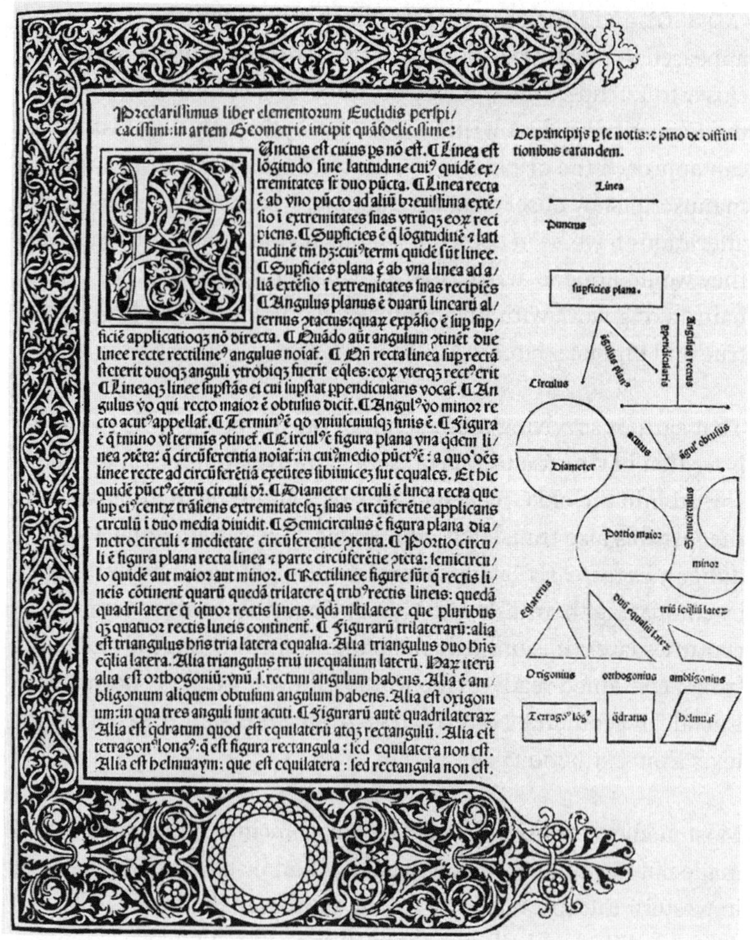
Preclarissimus liber elementorum Euclidis perspi/cacissimi: in artem Geometrie incipit quāfoelicissime:

Punctus est cuius ps nō est. ❡ Linea est lōgitudo sine latitudine cui⁹ quidē ex/tremitates sūt duo pūcta. ❡ Linea recta ē ab vno pūcto ad aliū breuissima extē/sio i extremitates suas vtrūq3 eo℞ reci/piens. ❡ Supficies ē q̄ lōgitudinē ⁊ lati/tudinē tm̄ h3: cui⁹ termi quidē sūt linee. ❡ Supficies plana ē ab vna linea ad a/liā extēsio i extremitates suas recipiēs ❡ Angulus planus ē duarū linearū al/ternus ɔtactus: qua℞ expāsio ē sup sup/ficiē applicatioq3 nō directa. ❡ Quādo aūt angulum ɔtinēt due linee recte rectiline⁹ angulus noiat. ❡ Qñ recta linea sup rectā steterit duoq3 anguli vtrobiq3 fuerit eq̄les: eo℞ vterq3 rect⁹ erit ❡ Lineaq3 linee supstās ei cui supstat ppendicularis vocat. ❡ An/gulus vo qui recto maior ē obtusus dicit. ❡ Angul⁹ vo minor re/cto acut⁹ appellat. ❡ Termin⁹ ē qd vniuscuiusq3 finis ē. ❡ Figura ē q̄ tmino vl termis ɔtinet. ❡ Circul⁹ ē figura plana vna q̄dem li/nea ɔtēta: q̄ circūferentia noiat: in cui⁹ medio pūct⁹ ē: a quo oēs linee recte ad circūferētiā exeūtes sibiinuice3 sut equales. Et hic quidē pūct⁹ cētrū circuli dr̄. ❡ Diameter circuli ē linea recta que sup ei⁹ cent℞ trāsiens extremitatesq3 suas circūferētie applicans circulū i duo media diuidit. ❡ Semicirculus ē figura plana dia/metro circuli ⁊ medietate circūferentie ɔtenta. ❡ Portio circu/li ē figura plana recta linea ⁊ parte circūferētie ɔtēta: semicircu/lo quidē aut maior aut minor. ❡ Rectilinee figure sūt q̄ rectis li/neis cōtinent quarū quedā trilatere q̄ trib⁹ rectis lineis: quedā quadrilatere q̄ q̄tuor rectis lineis. q̄dā mltilatere que pluribus q3 quatuor rectis lineis continent. ❡ Figurarū trilaterarū: alia est triangulus hñs tria latera equalia. Alia triangulus duo hñs eq̄lia latera. Alia triangulus triū inequalium laterū. Har℞ iterū alia est orthogoniū: vnū .s. rectum angulum habens. Alia ē am/bligonium aliquem obtusum angulum habens. Alia est oxigoni/um: in qua tres anguli sunt acuti. ❡ Figurarū autē quadrilatera℞ Alia est q̄dratum quod est equilaterū atq3 rectangulū. Alia est tetragon⁹ long⁹: q̄ est figura rectangula: sed equilatera non est. Alia est helmuaym: que est equilatera: sed rectangula non est.

3. The first page of the first printed edition of Euclid's *Elements*, 1482

no title page (because manuscript writers traditionally signed their names at the end of a text, not at the beginning), and it contains delicately painted illuminations (see Figure 3).

During the 16th century, printed editions followed each other rapidly, at first in Latin and Greek but then in several vernacular languages. Robert Recorde included most of the material from the first four books of the *Elements* in *The Pathway to Knowledg* in

1551, and some further and more difficult material from the later books in his last publication, *The Whetstone of Witte*, in 1557. The first full English translation of the *Elements* was published in a lavish edition by Henry Billingsley in 1570: it contains Dee's 'groundplat' and is also the earliest known English text to display the word 'mathematician' on the title page.

Over the next four centuries, there were many further translations and editions as editors adapted to the changing needs of the time. By the mid-20th century, the *Elements* was finally eased out of the school curriculum (though not its contents: schoolchildren still learn to construct triangles and bisect angles). It has not, however, disappeared from the public domain. A modern interactive web version is the most recent innovation in a very long tradition of translating and adapting the *Elements* for each new generation.

The *Elements* has been unique in its reach and longevity, but the story of its preservation is typical of that of many other Greek texts, including the *Arithmetica* of Diophantus, from which Fermat's Last Theorem arose. A similar story about early commentaries, translations into Arabic, later translations into Latin, and eventual print publication from surviving Greek sources can be told for most classical texts. There has been just one exception, the near miraculous rediscovery in the early 20th century of an otherwise lost text by Archimedes, faintly discernible below later writing and paintings on the pages of a Byzantine prayer book. Such finds are exceedingly rare, and serve to remind us yet again of how much mathematics of any culture has also been lost.

Preservation over distance

Despite the fragility of written documents, mathematics has been communicated not just over long periods of time, but sometimes over long distances, and sometimes both. We begin with a mystery.

Here is the beginning of a problem from an Old Babylonian tablet now in the British Museum (BM 13901):

> I summed the area and my square-side so that it was 0;45.

Using the technique that was warned against above, let us introduce algebraic notation for just long enough to see what the problem is about. If we let the side of a square be s then its area is s^2. The number 0;45 is a modern transcription that we may interpret as $\frac{45}{60}$ or $\frac{3}{4}$. Thus the statement can be written in modern terms as the equation $s^2 + s = \frac{3}{4}$. The Babylonian technique for finding the length of the square-side involved slicing and rearranging geometrical shapes; for the trained practitioner, this could be reduced to a series of brief instructions, a recipe guaranteed to give the answer.

Now consider this problem from a text on the subject of *Al-jabr wa'l-muqābala* ('Restitution and Balancing') composed by al-Khwārizmī in Baghdad around AD 825.

> A square and 21 units are equal to 10 roots.

Here the 'roots' are the square-roots of the given square, and so if we once again use modern notation we see that the problem can be written as $s^2 + 21 = 10s$. In other words, this is closely related to the Old Babylonian problem written down more than two and a half thousand years earlier. Further, al-Khwārizmī gave a very similar recipe for finding the answer. His text was so influential that it gave its name to the subject now known as algebra.

Is it coincidence that the same kind of problem with the same kind of solution reappeared so many centuries later in the same part of the world? There is no evidence at all for continuity down the years as we have for Euclid's *Elements*, certainly not within ancient or Islamic Iraq. We do, however, have evidence of ideas being carried from late Babylonian culture to India, and of mathematics later being transported in the other direction, from India to Baghdad. It is just possible that problems like those discussed here were part of that flow: we cannot say and can only speculate. It is worth rehearsing, however, what we know with more certainty.

From about 500 BC to 330 BC, ancient Iraq and north-west India were distant partners in the Persian empire, after which for a short time the same region came under the rule of Alexander the Great. Evidence for the absorption of Babylonian mathematics into India is circumstantial but fairly clear, especially in astronomical calculations: it can be seen in the Indian use of base 60 in measurements of time and angle, and in similar methods of calculating the length of daylight throughout the year. (In India, as in other early societies, correct time-keeping for ritual and other purposes was essential.) Later, there were translations into Sanskrit of Greek astronomical or astrological texts, so that the Greek 'chord', used in measuring astronomical altitude, became the basis of the Indian 'sine'. The dearth of early Indian texts prevents us from knowing what other knowledge must have passed eastwards, and no doubt in the other direction too: a few astronomical fragments from pre-Islamic Iran, for example, suggest the influence there of Sanskrit texts.

By the end of the 6th century AD (or even much earlier) there had been developed in parts of central India a system of writing numbers using just ten digits together with a system of place value. The importance of this can hardly be overstated. In modern parlance, it means that we can write any number of any size (or smallness) using just the ten symbols 0, 1, 2, 3, 4, 5, 6, 7, 8, 9. 'Place-value' means that '2' and '3' stand for different values in 200,003 and 302 because they are positioned differently. In both cases, the zeros serve as place-holders so that we do not mistake 200,003 for 23 or 302 for 32. Once this has been understood, the same few rules for addition and multiplication can be applied to numbers of any size. Of course, there have historically been many other ways of writing numbers, but all of them require the invention of more and more new symbols as the numbers get larger, and none is convenient for pencil and paper calculation: try adding a pair of numbers written in Roman numerals, xxxiv and xix for instance, without converting them into something more familiar.

The Indian or Hindu numerals, as they came to be called, were already known in parts of Cambodia, Indonesia, and Syria as early as the 7th century: they were highly praised by the Syrian bishop Severus Sebokht, for example. By AD 750 Islam had spread over the area of the old Persian empire (and beyond); and by 773 the Hindu numerals had arrived in Baghdad in astronomical treatises brought to the Caliph al-Manṣūr from India. Around 825, al-Khwārizmī, whom we have already met as a writer on *al-jabr*, wrote a text on the use of Indian numerals. The original is lost, but its contents can be recovered from later Latin translations. It taught first how to write the ten digits, in their Arabic rather than Sanskrit forms, with careful explanation of place-value and the correct use of zero; this was followed by instructions on adding and subtracting, doubling and halving, multiplication and division, some teaching on fractions including the sexagesimal kind, and directions for extracting square-roots. Al-Khwārizmī's text set the pattern for arithmetic texts for centuries: its outline can still be easily discerned in many 17th-century European texts even though the material was by then much expanded. For now, however, let us stay with the Indian numerals themselves or, as they had now become, the Hindu–Arabic numerals, as they continued to spread westward.

By the end of the 10th century, the numerals had been carried to Spain, at the other end of the Islamic world from India, and had acquired the western Arabic form that prefigured modern western numerals, rather than the eastern Arabic form still used in Arabic-speaking countries. And from Spain, they were slowly disseminated northwards into France and England. One of the myths about the numerals is that they were introduced into Christian Europe by a monk called Gerbert, later Pope Sylvester II, who had visited Spain before 970. It is true that Gerbert used the numerals on abacus counters, but on this slender evidence one can hardly give him credit for introducing them to the rest of Europe: we do not know whether he had learned the relevant methods of calculation or whether he merely used the numerals as decorative symbols; nor do we know how widely his abacus was known or used; and besides, there must have been other travellers to Spain who similarly brought back a little knowledge of the numerals to demonstrate to their friends. Knowledge of the numerals probably spread only slowly and in a piecemeal sort of way until their usefulness began to be better recognized.

We do know that astronomical tables from Spain, the Toledan tables, were adapted for Marseilles in 1140 and for London in 1150. The instructions for using the tables were translated from Arabic to Latin but the tables themselves were not: who would want to convert columns of two-digit figures measuring degrees, minutes, and seconds into clumsy Roman numerals? Just as astronomical tables had carried the Indian numerals to Baghdad, so they later brought them to northern Europe: for astronomers, the numerals were not just useful but crucial in order to make sense of other people's observations.

At a more mundane level, knowledge of the numerals and associated methods of calculation must also have spread westwards and northwards through trade. The Crusaders, for instance, would have encountered them from the late 11th century onwards. Unlike astronomical tables, however, records of buying and selling were ephemeral and have long since vanished.

By the 12th century, texts were being written specifically to explain the new numerals and the associated methods of calculation. One of them was Leonardo of Pisa's *Liber abaci* which circulated in Italy but not in northern Europe. In France and England, there were instead to be found Latin texts called 'algorisms', the name being a corruption of their opening words, 'Dixit Algorismi', meaning 'Thus spake al-Khwārizmī'. These texts, like al-Khwārizmī's original treatise, taught how to write the numerals and how to carry out basic arithmetic with them. A particularly charming one, known as the 'Carmen de algorismo', was composed in verse by Alexander de Ville Dieu from northern France. The opening lines are, in translation:

> This present art is called algorismus, in which
> We make use of twice-five Indian figures:
> 0.9.8.7.6.5.4.3.2.1.

Alexander went on to explain how the position of each numeral mattered:

> If you put any of these in the first place,
> It signifies simply itself: if in the second,
> Itself tenfold.

Despite their obvious advantages, the uptake of the numerals was slow, not, as is sometimes suggested, because of their Oriental and non-Christian origins, but because for everyday use the old Roman system together with calculations done on fingers or counting boards served well enough. Besides, not everyone found the new numerals easy to learn: as late as the 14th or 15th century a monk in the Benedictine monastery of Cavenso in Italy numbered some of his chapters from the 30th onwards as XXX, XXX1, 302, 303, 304, ... Eventually, however, the Hindu-Arabic numerals superseded all others, and once they had been taken west to America had almost completed their circumnavigation of the world.

There are other stories that could be told about the way mathematics has been disseminated over long distances. Traditional Chinese mathematics, for instance, was taken up by all China's immediate neighbours, and there were no doubt exchanges with India as well, but none with the West until the Jesuits arrived in the 17th century, bearing the *Elements* of Euclid. Such movements have continued in more recent times: in the 19th century, European mathematics was carried outwards from its heartlands in France and Germany to the peripheries of Europe, the Balkans at one end and Britain at the other, and then to the United States, and eventually to every country in the world. Such dissemination is typical of the modern era, but in mathematics, ideas have been travelling for a very long time.

Not forgetting people

In this chapter I have described how some of the mathematics of the past has survived, in however fragmented a form, over long periods of time and sometimes over long distances too. I have tried to be careful, however, with language. A common word for the passing on of mathematical ideas is 'transmission', but I dislike it: apart from the connotation of radio masts, it suggests that the originators were deliberately aiming their ideas and discoveries at future generations. This has rarely been the case. For the most part, mathematics is written down for one's own use or for one's immediate contemporaries, and its survival much beyond that is largely a matter of circumstance. I have also tried to avoid speaking of ideas simply spreading, as though they were garden weeds with a power of their own.

On the contrary, every mathematical exchange, large or small, is brought about by human agency. Behind the long-running stories outlined above lie innumerable tiny interactions and transactions. We have already glimpsed some of them: Indian envoys presenting themselves to the caliph in Baghdad; a Byzantine scribe copying a manuscript he may barely have understood;

Florentine traders haggling in the markets of Alexandria; a library-keeper in that same city of Alexandria a millennium earlier carefully listing the scrolls in his charge, and perhaps, like John Leland later, distraught at the thought of their destruction; Fermat sending his letters in false hope to Wallis in Oxford; Wiles delivering first news of his proof in a lecture, and news of its eventual correction by email. Mathematical ideas move around only because people think about them, discuss them with others, write them down, and preserve relevant documents. Without people, there is no dissemination of mathematical ideas at all.

Chapter 4
Learning mathematics

A fact easy to overlook is that the largest group of people in modern society doing mathematics is composed not of adults but of schoolchildren. A young person anywhere in the world who is fortunate enough to have access to education is likely to spend a significant amount of time learning mathematics; in the developed countries, this is likely to amount to two or three hours of each school week for ten years or more.

In the light of that, it is a little surprising to remember that the inclusion of mathematics in the school curriculum is a modern phenomenon. Around 1630, for example, John Wallis, later Savilian Professor of Geometry at Oxford, learned arithmetic neither at school nor at Cambridge but from his younger brother studying to go into trade; 30 years later, the highly intelligent and literate Samuel Pepys, also educated at Cambridge, and a member of the Navy Board, struggled to learn his multiplication tables. Nevertheless, passing on mathematical knowledge to at least a few of the next generation has been regarded as an important task in most civilized societies.

A study of what has been taught, and how, tells us a good deal about what aspects of mathematics have been regarded as relevant, and for what purposes. In this chapter we will examine two case studies for which we have relatively good documentation:

a schoolroom at Nippur in southern Iraq some time before 1740 BC, and another at Greenrow Academy in Cumbria in the north of England shortly after AD 1800.

A Babylonian schoolroom

The ancient city of Nippur, situated in the marshlands of the Euphrates about halfway between the modern cities of Baghdad and Basra, was an important religious centre, built around a temple complex dedicated to the god Enlil. Like the abbeys and monasteries of medieval Europe later, Babylonian temples received substantial offerings, and controlled land and labour, and so needed trained scribes who could handle written accounts and calculations. Children destined for the profession, which usually ran in families, probably began their training early.

A small mud-brick house in Nippur, now known as House F, appears to have been one of perhaps several scribal schools in the city. Close to a temple dedicated to the goddess Inana, House F was first built some time after 1900 BC and was used as a school shortly before 1740 BC. Like all mud-brick structures, it required regular upkeep and after it ceased to be used as a school it was rebuilt for the fourth or fifth time. At this point, the builders made good use of hundreds of discarded school tablets which they incorporated into the floor, walls, and furniture of the new house. Other partially destroyed tablets have been found mixed with large quantities of unused clay in recycling bins.

When used as a school, House F was divided into three or four inner rooms and two courtyards, the latter containing benches and the recycling bins. Unfortunately, we do not know the names or ages of the students, of whom there may have been no more than one or two at a time, nor do we know how often or for how long they occupied the courtyard benches. Remarkably, however, their method of using their tablets has enabled cuneiformists to reconstruct their curriculum.

Many of the tablets from House F are flat on one side (the obverse) and slightly rounded on the other (the reverse). The obverse contains on the left a model text written by a teacher, copied on the right by the student. The rounded reverse of the tablet, however, contains longer passages of material learned earlier, rewritten for further practice or perhaps as a test of memory. From some 1,500 Nippur tablets of this kind, each containing 'earlier' and 'later' material, Niek Veldhuis in the 1990s was able to discern a consistent order in the elementary curriculum, starting with basic writing techniques and ending with the beginnings of literary Sumerian. Applying the same methodology to about 250 similar tablets from House F, Eleanor Robson was able to do the same for the House F curriculum, and so to discover the place of mathematics within it.

The student's first steps were to learn the correct technique for writing cuneiform signs and to combine them to form personal names. Then they acquired written vocabulary through lists of words, beginning with trees and wooden objects, then reeds, vessels, leather and metal objects; animals and meats; stones, plants, fish, birds, and garments; and so on. Some mathematical vocabulary was already introduced here, with measures of capacity for boats; of weight for trees and stones; and of lengths for reed measuring-rods. Further metrological units also appeared in dedicated lists of weights and measures later.

Next, the student was expected to learn by heart lists of inverses (pairs of numbers that multiply to 60) and more than 20 standard multiplication tables. A list of inverses, for example, might begin

2	30
3	20
4	15
5	12
6	10
8	7 30
9	6 40
10	6
12	5

(In sexagesimal arithmetic, which we still use for hours, minutes, and seconds, 7 30 is equivalent to $7\frac{1}{2}$, and 6 40 to $6\frac{2}{3}$.) Multiplication tables required considerable feats of memory. The multiplication table for 16 40, for example, began

1	16 40
2	33 20
3	50
4	1 06 40
5	1 23 20

It has been estimated that it could take up to a year to learn the full set of tables alongside other school exercises. At this stage, students also started to write complete sentences in Sumerian, some of them containing the metrological units learned earlier.

Only after all this, as they also learned more advanced Sumerian, did students begin to carry out their own calculations of reciprocals or inverses not in the standard tables. One of the few 'advanced' tablets from House F contains some of the calculations used to find the inverse of 17 46 40 (answer: 3 22 30). These are written on the same tablet as an extract from a literary work known as 'The supervisor's advice to a younger scribe', which includes some moralistic lines based on the supervisor's memory of himself as a young student:

Like a springing reed, I leapt up and put myself to work.
I did not depart from my teacher's instructions;
I did not start doing things on my own initiative.
My mentor was delighted with my work on the assignment.

Most of the advanced texts from House F are not mathematical but, like 'The supervisor's advice', literary compositions. Many, however, contain references to the uses of both literacy and numeracy in the just administration of society. Lines from a hymn to Nisaba, the patron goddess of scribes, praise her for bestowing her gifts upon the king:

A 1-rod reed and a measuring rope of lapis lazuli,
A yardstick, and a writing board which gives wisdom.

A Cumbrian schoolroom

Greenrow Academy was founded in 1780 by John Drape, at Silloth on the north-west coast of England, just a few miles south of the Scottish border. Like the school in House F at Nippur, Greenrow Academy was something of a family enterprise. Drape's father, known as John Draper, had previously run a school in Whitehaven, 30 miles south on the same coast. The Whitehaven school had emphasized subjects relevant to 'trade and seamanship' and Draper had published two textbooks for the use of his pupils: *The Young Student's Pocket Companion, or Arithmetic, Geometry, Trigonometry, and Mensuration, Calculated for the Improvement of Youth at School* (1772) and *The Navigator's veni-mecum: or a Complete System of the Art of Navigation* (1773). When Draper died in 1776, his son John inherited his books, his mathematical instruments, and some property, enabling him to establish Greenrow Academy a few years later. After Drape himself died in 1795, the school passed into the care of another family member, Joseph Saul, a relative of Drape's wife, who remained in charge of it for nearly 50 years. The curriculum was broadened to include

Greek, Spanish, and scriptural studies, but Greenrow Academy, like its parent school in Whitehaven, continued to place a strong emphasis on mathematical studies.

The school attracted boys not only from the local area but from elsewhere in England and even from overseas. Nine-year-olds, and at one time even a six-year-old, could be registered, but young men in their early twenties were also sometimes educated there. Most of the pupils, however, were aged about 14 or 15. Records for 1809 show that one of the youngest pupils, Rowland Cowper (aged 11), and one of the oldest, James Irving (aged 23), followed the same basic curriculum of English, writing, and arithmetic. Most of the other boys also studied drawing and either French or Latin together with a wide range of mathematical topics. The curriculum followed by John Coleman (aged 15) was typical: English, French, writing, drawing, arithmetic, geometry, trigonometry, mensuration, surveying, book-keeping, spherics, astronomy, mechanics, algebra, and Euclid. Further mathematical subjects on offer were dialling (the construction of sun-dials), gauging, and fortification, while George Peat (aged 16), who seems to have been exceptionally able, also took lessons on conic sections and fluxions (the Newtonian version of the calculus).

We are fortunate, however, to have more from Greenrow than mere lists of topics. Before his death in 2005, the mathematics educator John Hersee collected over 200 mathematical copy books written by pupils in schools throughout England and Wales between 1704 and 1907. These were not exercise books in the modern sense. Pupils did not waste precious paper practising similar questions over and over again; instead they inscribed carefully copied examples of standard problems, thus creating for themselves a collection of worked examples which they could carry with them into later life. Many of the examples were taken from popular textbooks of the time, in particular from the long-running *The Tutor's Assistant* of Francis Walkingame (first published in

1751), but many others must have been invented for their pupils by teachers themselves.

The Hersee collection includes five mathematical workbooks by Robert Smith from the years 1832 and 1833 (see Figure 4). Over these two years Robert filled almost 1,700 pages with mathematical examples, so that we have a very detailed picture of

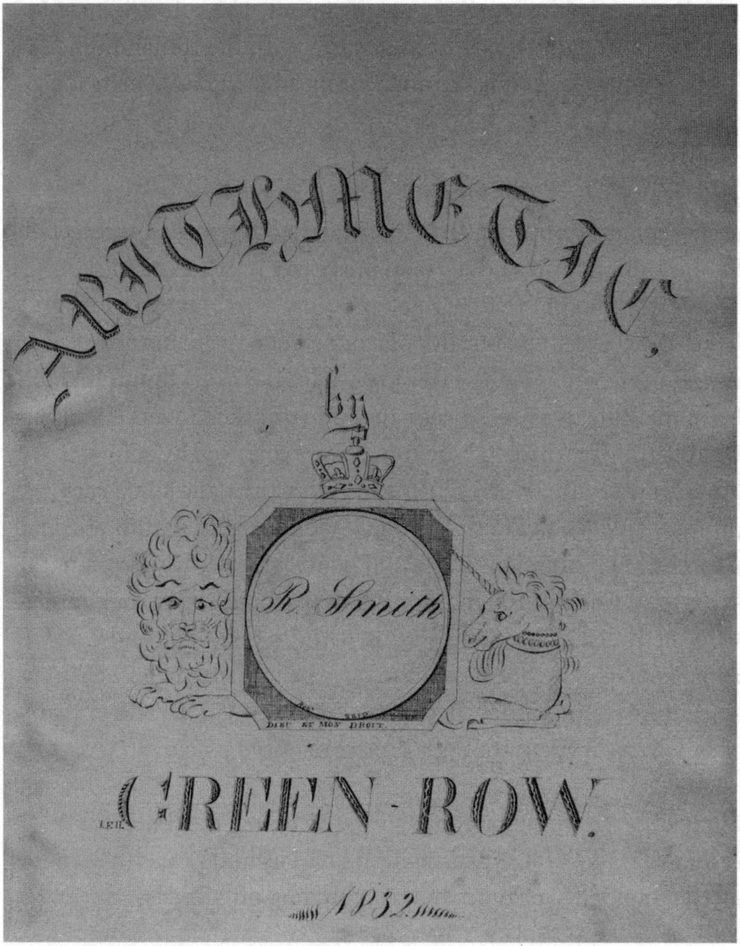

4. The front page of Robert Smith's arithmetic book, Greenrow Academy, 1832

what he studied. These books were not the first that Robert wrote because he had already advanced beyond the elementary operations of addition, subtraction, multiplication, and division. The earliest surviving book, for 1832, begins with the Rule of Three. This was the rule that enabled countless generations of students to answer questions like: A men dig a ditch in B days, how long would it take C men to do the same job? The rule is so named because there are three known quantities (A, B, C) from which a fourth (the answer) must be found. The rule originated in India and probably travelled westwards with the Indian numerals: it was ubiquitous in Islamic and European arithmetic texts for centuries.

The Rule of Three was taught by rote: like his Babylonian predecessors, a 19th-century English schoolboy was not expected to 'start doing things on my own initiative'. In the example above, he would be taught that he must multiply B by A and divide by C to get the correct answer. But of course, there were always variations to catch out the unwary student: Robert Smith had to learn the Rule of Three Direct, the Rule of Three Inverse, and the Double Rule of Three. These topics were followed by, among others, Barter, Interest, the Rule of Fellowship (the sharing of profits), Vulgar Fractions, Decimal Fractions, and Arithmetic and Geometric Progressions. A second book, apparently written in the same year, works through a similar list of topics, again beginning with the Rule of Three and ending with Progressions and Duodecimals. The books seem to have been written consecutively, since Robert himself labelled them Vol. I and Vol. II, and it is not clear why he worked through similar material twice.

Many of his examples are taken from Walkingame; here, for instance, is one of just two examples on Permutations (the other is on the number of changes that can be rung on 12 bells):

> A young person, coming to town for the convenience of a good library, made a bargain with the person with whom he lodged, to

> give him £40 for his board and lodging, during so long a time as he could place the family (consisting of 6 persons besides himself) in different positions, every day at dinner. How long might he stay for his £40?

Robert wrote the correct solution ($1 \times 2 \times 3 \times 4 \times 5 \times 6 \times 7 = 5040$ days) immediately after the question, but then, following Walkingame quite closely at this point, moved on immediately to vulgar fractions.

Robert's two arithmetic books for 1832 between them contain almost 900 pages. In addition, he filled almost 500 pages more in a third book headed 'Geometry Trigonometry Mensuration and Surveying', which contains some of the beautifully drawn and painted sketches that seem to have been encouraged at Greenrow (see Figure 5).

The next book, with the title page 'Arithmetic by Robert Smith Green-Row 1833', is on 'Practical questions upon common rules'. The questions known as Bills of Parcels are of particular interest because pupils often substituted their own names and dates for those given by Walkingame. Thus Robert's first Bill begins:

> Green Row July 13th 1832
> Mr Thos Nash
> Bought of Robert S Smith

8 Pair of worsted stockings	at 4s 6d per pair	£1 16s 0d
5 Pair of thread ditto	at 3s 2d	15s 10d

Further dates on further bills follow through July 1832 and into August, suggesting that Robert might have been writing this book in 1832, and did not begin it but only finished it in 1833, the date on the title page. The name of Thomas Nash is elsewhere inscribed at the end of Robert's first book, along with that of one

5. A problem in trigonometry, illustrated and answered by Robert Smith, Greenrow Academy, 1832

Robert Reid, suggesting that they might have been his teachers; Robert Reid makes another appearance here, charged with

18 Yards of fine lace	at £0 12s 3d per yard	£11 0s 6d
5 Pair of fine kid gloves	at 2s 3d per pair	11s 3d

and so on.

The second book completed in 1833 was on 'Mensuration of Solids'. It includes sophisticated calculations of the volumes and surface areas of the five regular solids (tetrahedron, cube, octahedron, dodecahedron, icosahedron) but also calculations typical of those used by bricklayers, masons, carpenters, slaters, painters, glaziers, plumbers, and others, with the appropriate units used by each. Robert learned that painters, for example, estimate the areas of 'wainscottings, doors, window shutters' by the square yard but that 'deductions must be always made, for fireplaces and other openings'.

Unfortunately, we do not know how old Robert was when he did all this, but we can see that his years at Greenrow gave him a mathematical education that was both theoretical and thoroughly practical.

Girls

I have hesitated to include a section that treats a group of people who constitute half of humanity as though they were a minority, but there is no getting away from the fact that for most of history in most societies it has not been thought necessary, or indeed appropriate, to educate girls, and certainly not in subjects like mathematics or science. It is not surprising, therefore, that there have been very few women mathematicians of note, just as until recent centuries there have been few women writers, lawyers, or doctors. This state of affairs must have left countless thousands of intelligent women somewhat frustrated. Nevertheless, there have from time to time been some who were given, or who created for themselves, opportunities for mathematical education.

One group of such women were those with enough wealth or leisure to indulge in whatever studies they chose. An early example was the Chinese Empress Dèng 鄧, who at the end of the 1st century AD took lessons in *suàn shù*. Unusually for this period, her teacher was also a woman, by the name of Bān Zhāo 班昭. Much later, in the 1640s, Princess Elisabeth of Bohemia and Queen Christina of Sweden both took lessons from Descartes, though they were perhaps more interested in his philosophy than in his mathematics. A century later, Europe's most prolific mathematician, Leonhard Euler, wrote over 200 letters on mathematical and scientific subjects to the Princess of Anhalt-Dessau, niece of Frederick the Great of Prussia. The letters, published in French, Russian, and German, and later in English as *Letters to a German Princess*, remain in print to this day.

A more common route into mathematics for ordinary women, however, was to be taught by a father, husband, or brother. In the 19th century BC, for example, there were two women scribes in the Babylonian town of Sippur, the sisters Inana-amağa and Niğ-Nanna. It seems more than likely that they learned the profession from their father, Abba-ţābum, also a scribe. Two millennia later, the Empress Dèng and her brothers received their earliest education from their father, even though their mother, it seems, thought this a waste of time for a girl. The Empress Dèng's later teacher, Bān Zhāo, was the sister of the scholar Bān Gù 班固, whose work she understood sufficiently well to complete after his death, including a treatise on astrology. Perhaps the most famous father–daughter pair in mathematics is that of Theon and Hypatia in late 4th-century Alexandria; from Hypatia herself, however, we have no writings, only secondary accounts of her life and death around which much legend has accumulated.

Teaching of girls within the family continued into the early modern period. John Aubrey, writing in the 1670s about his former friend Edward Davenant, vicar of Gillingham in Dorset, noted his love of mathematics, though 'being a Divine he was unwilling to print, because the world should not know how he spent a greater part of his time'. Davenant had taught algebra not only to Aubrey himself, but also to his own daughters:

> He was very ready to teach and instruct. He did me the favour to informe me first in Algebra. His daughters were Algebrists.

As it happens, we know what Edward Davenant taught his eldest daughter, Anne, in the way of algebra because in 1659 Aubrey, avid recorder of all human affairs, copied out the contents of her notebook. Anne was born before 1632 (the date of birth of her younger sister Katherine) and married Anthony Ettrick in 1650, so was probably being trained as an 'Algebrist' some time in the early 1640s. Aubrey's copy of her work is headed

> This algebra I transcribed from ye MS of M^{ris} Anne Ettrick the eldest daughter of Dr Davenant who is a very good Logist.

The early problems in Anne's notebook, and the Latin in which she wrote them, are typically those of a young beginner. In one of them, for example, some girls are taking a walk when a young man comes along. 'Greetings', he says (in Latin), 'you twelve maidens'. To which one of the girls replies immediately (and also in Latin): 'If we were multiplied by five, we would be as many more than twelve as we are now fewer than twelve.' Question to the reader: how many girls were there? Several pages later, we find Anne working an example whose form and solution had been proposed by al-Khwārizmī in Baghdad eight centuries earlier: what number multiplied by 6 and added to 16 makes its own square (in modern notation $6x + 16 = x^2$). Finally, towards the end of the notebook, both the Latin and the mathematics become more mature. The penultimate problem comes straight from the *Arithmetica* of Diophantus: to divide 370 into two cubes, whose roots are whole numbers that add to 10. As Anne was able to show, the answer is 7^3 plus 3^3. The numbers in this question have been carefully chosen to give an easy solution, but it becomes impossible if 370 is replaced by a perfect cube, as Fermat, at much the same time but far away in Toulouse, was just discovering.

Well into the 18th century, girls were likely to be taught mathematics only if, like the Empress Dèng or Anne Davenant, they had the advantages of social status or indulgent parents. Sophie Germain, one of the key people to make progress on Fermat's Last Theorem, benefited from both. Born into a wealthy and educated family in Paris in 1776, she was just 13 when the French Revolution broke out. Confined to her home, she amused herself in her father's library and discovered mathematics, a subject her parents did not at first think suitable for her but they later relented in the face of her determination. When she was 18, she managed to obtain lecture notes from the newly opened École Polytechnique and, although she was not allowed to attend lectures herself, she submitted work under the pseudonym Monsieur le Blanc to one of the greatest of the École's teachers, Joseph-Louis Lagrange. Four years later, she corresponded with the equally great German mathematician Carl Friedrich Gauss, again under the pseudonym le Blanc. To the credit of both Lagrange and Gauss, they continued to admire her mathematics and her courage even after they discovered her true identity. Sophie struggled against the odds for most of her life: she had never had the kind of education that a boy of equal talent might have had, and her work was often marred by errors and incompleteness. She never held any official post. Nevertheless, after her death in 1831 Gauss remarked that she would have been worthy of an honorary degree at Göttingen, by then one of the key mathematical centres of Europe.

Articles or posters about 'Women in mathematics' invariably feature Hypatia and Sophie Germain, unfortunately not because they were typical of their time and place but because they were not. Unsung women like Bān Zhāo or Anne Davenant are on the whole much more representative of women's experience of mathematics and mathematical education.

By the 19th century, the situation for girls in western Europe slowly improved as they started to benefit in larger numbers from elementary school education. Copy books by girls are in the minority in the Hersee collection, but those that we have give us some valuable insights into the kind of mathematics taught to girls in a variety of schools in England and Wales.

In 1831, the year before Robert Smith at Greenrow began creating the books discussed above, Eleanor Alexander, at Fairwater School in a valley north of Newport in South Wales, worked on Reduction ('Bring £30 1s $1\frac{1}{4}$ into farthings') and the Rule of Three ('If 7 yards of cloth cost £3 10s what will 65 yards cost?'). Her entire copy book of 127 pages consists of just these two types of question. Three years later, beginning in October 1834, Ann Weetman at Appleton-le-Moors, near York, began to work her way through Walkingame's *The Tutor's Assistant* (see Figure 6). Her early entries are dated, so we know that she spent about ten days on Simple Addition but a whole month on Multiplication. After Christmas she worked on Compound Addition (of money, cloth measure, land measure, beer and ale measure, and so on) and by the end of March she reached Bills of Parcels. In her case, the 8 pairs of worsted stockings went to Mrs Wm G. Atkinson (her teacher?) while Mr Henry Weetman (her father? her brother?) bought 15 yards of satin. By April 1837 she had reached Practice, a method that relied on knowing fractions of standard weights and measures (see Figure 7). Her book ends after 250 pages on 10 May 1837, by which time she had worked her way through Walkingame as far as Compound Interest, considerably less in three years than Robert Smith had completed in a few months, but still a respectable amount of mathematics.

Twenty years later, Elisabeth Attersall in Stainfield in Lincolnshire also worked through Walkingame, from Compound Addition to the Rule of Three ('Gave £1 1s 8d for 3lb of coffee, what what [sic] must be given for 29lb 4oz.'). In her case, the 8 pairs of worsted stockings went to Mrs Chappell on 22 October 1850. Miss I. Norman at Mr Ingleson's Dorset Street Academy in Hulme, Manchester, in 1861, however, was served a slightly

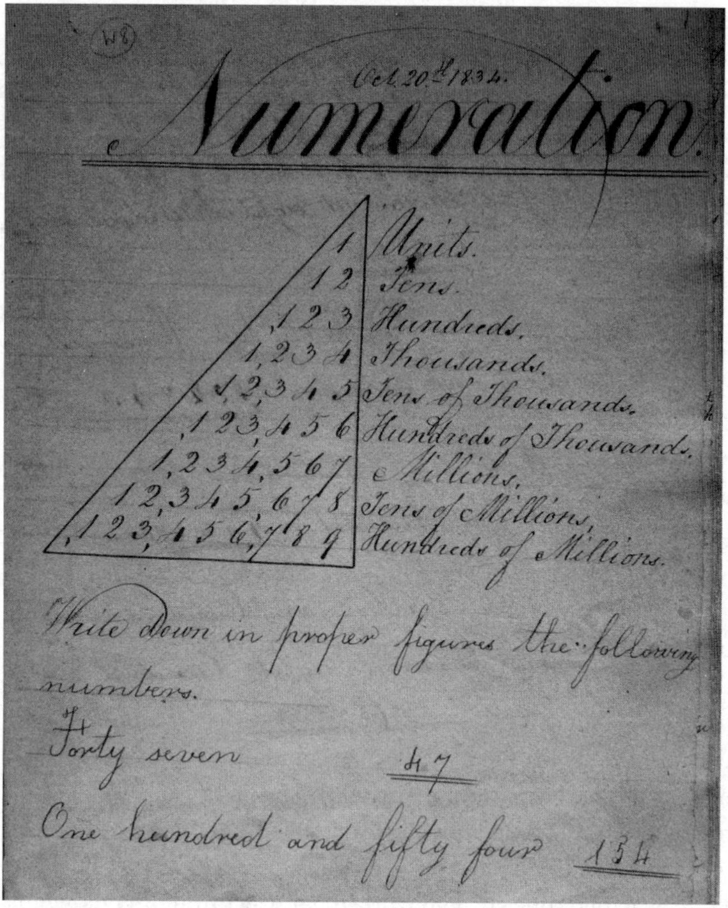

Oct 20th 1834.
Numeration.
Units.
Tens.
Hundreds.
Thousands.
Tens of Thousands.
Hundreds of Thousands.
Millions.
Tens of Millions.
Hundreds of Millions.
Write down in proper figures the following numbers.
Forty seven
47
One hundred and fifty four
154

6. The first page of Ann Weetman's exercise book, dated 'Oct 20th 1834'

different diet. Her copy book was specially printed with the name of the school, on pale blue paper with margins ruled in double red lines. On the first page, Miss Norman wrote 'Progressive arithmetic' but sadly she did not progress very far: every one of the book's 60 pages is filled with multiplication or division of pounds, shillings, and pence ('What must I pay for 4767 yards of cloth at $7\frac{8}{9}$d per yard?'). Elizabeth Dawson at Carshield School in Northumberland a year later did a great deal of work on the Rule of Three and a great deal more on Practice; to find 'the value of

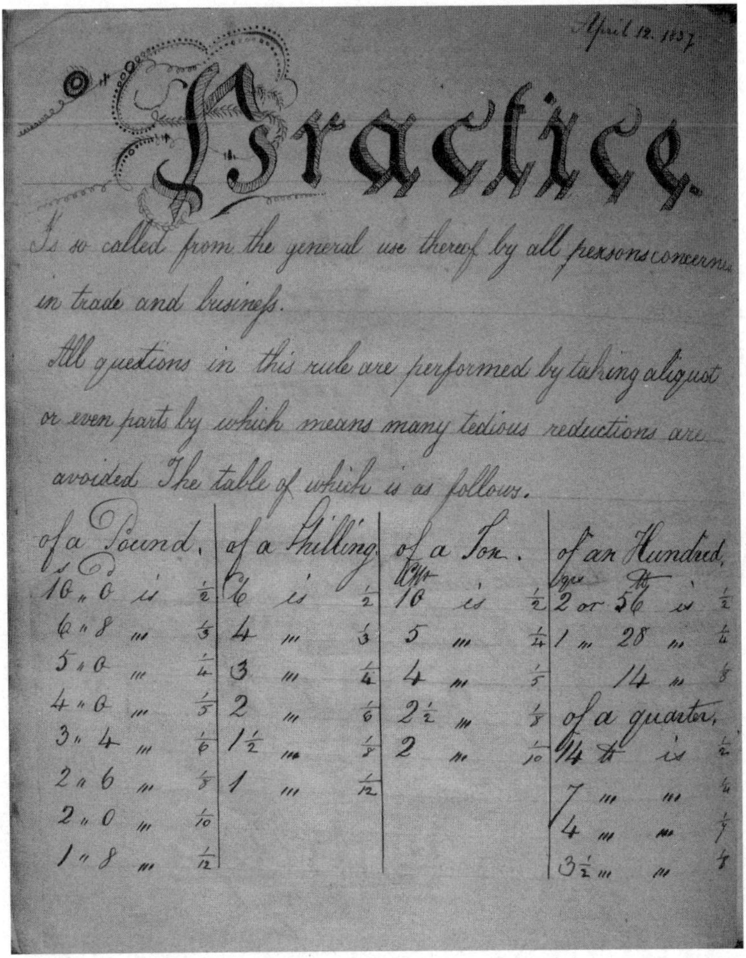
April 12. 1837

Practice

Is so called from the general use thereof by all persons concern[ed] in trade and business.

All questions in this rule are performed by taking aliquot or even parts by which means many tedious reductions are avoided. The table of which is as follows.

of a Pound.			of a Shilling			of a Ton.			of an Hundred,		
s d						Cwt			qrs lb		
10 " 0	is	$\frac{1}{2}$	6	is	$\frac{1}{2}$	10	is	$\frac{1}{2}$	2 or 56	is	$\frac{1}{2}$
6 " 8	"	$\frac{1}{3}$	4	"	$\frac{1}{3}$	5	"	$\frac{1}{4}$	1 " 28	"	$\frac{1}{4}$
5 " 0	"	$\frac{1}{4}$	3	"	$\frac{1}{4}$	4	"	$\frac{1}{5}$	14	"	$\frac{1}{8}$
4 " 0	"	$\frac{1}{5}$	2	"	$\frac{1}{6}$	$2\frac{1}{2}$	"	$\frac{1}{8}$	of a quarter,		
3 " 4	"	$\frac{1}{6}$	$1\frac{1}{2}$	"	$\frac{1}{8}$	2	"	$\frac{1}{10}$	14 lb	is	$\frac{1}{2}$
2 " 6	"	$\frac{1}{8}$	1	"	$\frac{1}{12}$				7 "	"	$\frac{1}{4}$
2 " 0	"	$\frac{1}{10}$							4 "	"	$\frac{1}{7}$
1 " 8	"	$\frac{1}{12}$							$3\frac{1}{2}$ "	"	$\frac{1}{8}$

7. One of the final pages of Ann Weetman's exercise book, dated 'April 12 1837'

7234 yards at 6s 8d', for example, she used a fact known to every British schoolchild before 1971, that 6s 8d was £$\frac{1}{3}$. It was rather less easy, though, for her to find the cost of 65 feet $3\frac{1}{4}$ inches at 3s $7\frac{3}{4}$d per foot.

From April 1866, Isabella Lund, a pupil at the Grammar School in Bolton-le-Sands in Lancashire (founded originally only for boys), spent just over a year progressing from Simple Addition to the Rule of Three. A year later, Miss G. Jones at Ribston Hall Grammar School for Girls in Gloucester gradually filled out 20 pages of invoices; her 8 pairs of worsted stockings went to a Miss Jenkins in July 1868.

The girls' copy books selected above are some for which we happen to know the name of the writer and the school and the date; without further research we cannot assume they are representative. They do suggest, however, that mathematical education for girls had a strong practical emphasis (no Euclid here); further, by modern standards, the pace was sometimes agonizingly slow and repetitive. Nevertheless, the girls who wrote these books were both literate and numerate, particularly compared with their counterparts in previous generations.

To go beyond elementary mathematics to university education, however, still required a particular strength of character. We will end this section with a comparison of two women who did manage to reach the higher reaches of their country's education systems, Flora Philip in Scotland and Florentia Fountoukli in Greece.

Flora Philip was one of the first women graduates of Edinburgh University, in 1893, but she had already joined the Edinburgh Mathematical Society seven years earlier. Most of her education in higher mathematics was gained in fact not from the university but through the Edinburgh Association for the University Education of Women. The Association had been established in 1867 to provide education beyond school level for women, parallel to that provided by the university for men. Mathematics was included in the Association's lecture courses early on, though not without some opposition from those who considered it 'altogether outside the domain of a lady's thought'. The aim was to teach the same mathematics as was taught at the university, but since many of the women were ill-prepared by their elementary schooling, the level reached was never as high as in the university courses. Nevertheless, they were taught Euclid, algebra, trigonometry, and conic sections. The number of women who took the courses was often very low but, as one of the lecturers reported, 'The zeal and industry of the Class doubly compensates for the smallness of its numbers'. Later, a more advanced senior course was introduced, from which Flora Philip graduated successfully in 1886, the year she joined the Edinburgh Mathematical Society. By the time she was awarded her degree from the university in 1893, she had already been teaching for some time at St George's School for Girls, a school founded by the Association. In that same year, however, she married, and thereafter withdrew both from academic life and from the Edinburgh Mathematical Society.

The career of Florentia Fountoukli, born in 1869 in Athens, ran parallel in several ways to Flora's. While Flora was studying mathematics with the Edinburgh Association in the 1880s, Florentia Fountoukli was earning her schoolteacher's diploma from the Arsakeion Normal School for girls in Athens. Afterwards the Board of the Arsakeion Schools allowed Florentia funds to study pedagogy in Berlin for a year. She then requested an extension in order to take a degree in Zürich, in mathematics, but the Board refused. (Her brother Michael, on the other hand, did become a mathematician and later worked in Hamburg.) Florentia returned to teach in the Arsakeion School in Corfu, during the same years that Flora was teaching at St George's. In 1892, just as Flora matriculated into the University of Edinburgh, Florentia matriculated into the mathematics department of the University of Athens, the first woman to do so. Unlike Flora, however, she does not appear to have graduated. Instead she continued to teach, in a school for girls in Athens, which she and her friend Irene Prinari founded and ran themselves. By 1899, she was signing herself Fountoukli-Spinelli which suggests that she may have married Loudovikos Spinelli, another teacher, but the facts are not clear. Unfortunately, by the late 1890s, before she was even 30, her health began to fail and she went to live in Italy, where she died in 1915.

Both Flora and Florentia had to struggle to get the kind of education they wanted. Nevertheless, the universities of Edinburgh and Athens were ahead of some others. Cambridge University did not admit women as full members until 1947.

Autodidacts

Until two centuries ago, only a small number of girls anywhere in the world received any kind of mathematical education. But even for boys, compulsory education in mathematics has been a relatively recent phenomenon. In 17th-century England, as we saw with Wallis and Pepys, it was possible to be educated all the way through school and university without learning much at all in the way of mathematics. Those with a particular aptitude or liking for the subject were therefore often largely self-taught. This was the case for Fermat, who learned some of the most advanced mathematics of his day from the books owned by the father of his friend Etienne d'Espagnet in Bordeaux. It was also the case for Isaac Newton, one of the greatest mathematicians of the 17th century. Newton may have learned some elementary mathematics at his grammar school in Grantham in Lincolnshire, but he learned very much more through his own reading as a student at Cambridge in the 1660s. Many years later, he described to a friend how he had read Descartes' *Géométrie*, republished in Latin just a few years earlier. Many people will recognize the difficulties of reading a strange new mathematical text; perhaps rather fewer will have emulated Newton's stubborn and self-motivated persistence.

> He bought Descartes's Geometry & read it by himself. When he was got over 2 or 3 pages he could understand no farther, then he began again & got 3 or 4 pages farther till he came to another difficult place, then he began again & advanced farther & continued so doing till he made himself Master of the whole.

We know from Newton's surviving manuscripts that he proceeded in a similar way with other contemporary texts, and that he worked on the material he found in them to create mathematics that went far beyond that of any of his predecessors.

In the 17th century, and some way into the 18th, a person sufficiently motivated could still read and learn from almost the entire range of available mathematical literature. Even at the beginning of the 19th century, Sophie Germain managed to teach herself some of the most advanced mathematics of her day, but she belonged to almost the last generation for whom this was feasible. By the 20th century, it was barely possible except, perhaps, for prodigies with quite exceptional mathematical instincts, like the self-taught Indian mathematician Ramanujan. Andrew Wiles was certainly not self-taught; he went through the many years of formal education that are now needed to initiate even the most mathematically gifted into some of the problems, techniques, and conventions of the discipline. The days of 'amateur' mathematics when almost anyone could have a go at a cutting-edge problem like Fermat's Last Theorem are long past.

Nevertheless, from time to time writers continue to create fictional accounts of individuals who manage to learn mathematics from the writings of another person, well enough to understand and extend their work. One such story is *Mrs Einstein* by Anna McGrail, another is the more recent play *Proof* by David Auburn. In both, the heroines are the daughters of mathematicians (a trope we have already seen in real life) who manage to educate themselves to very high levels in their fathers' work. Unfortunately, the reality of modern mathematics is that such feats are now extremely implausible.

Why learn mathematics at all?

Given the enormous amount of human energy that over the centuries has gone into teaching or learning mathematics, it seems a little perverse to ask 'Why?' The answers to the question, however, have varied considerably over time. Sumerian texts of the 2nd millennium BC made it clear that literacy and numeracy were essential to the just administration of society, though that may have seemed a somewhat distant ideal to the boys on the cramped courtyard benches of House F.

Two thousand years later, boys of a similar age educated in the abacus schools of 13th-century Italy learned, like their earlier Babylonian counterparts, to handle numbers, weights, and measures, but for different reasons: not for the good of society as a whole but so that they as individuals would be better fitted for the commercial ventures they were expected to engage in. The value of mathematical skills to the individual is seen again in Robert Recorde's preface to *The Pathway to Knowledg* with its long list of specific crafts and occupations that require a knowledge of geometry.

In Recorde's writings, however, we also get a glimpse of yet another reason for studying mathematics: to whet the wits, that is, to sharpen the mind. Recorde was not the first to suggest this. Some mathematical puzzle questions ascribed to the 8th-century teacher Alcuin were entitled 'Propositions of Alcuin for sharpening the young' (*'Propositiones Alcuini ad acuendos juvenes'*). The idea that mathematics should be learned, like Latin or Greek, to improve one's mind has continued ever since. After all, the mathematics required for ordinary daily life, basically time-keeping and accounting, has probably been acquired by most people by the end of childhood. Few adults ever need to use Pythagoras' Theorem, or solve quadratic equations, or bisect an angle, but almost all were once taught to do so. It can be argued (and I would argue) that learning a foreign language or studying history does just as much to encourage the development of memory, reasoning, and analysis, but such subjects have never acquired the prestige of mathematics, and at present are optional rather than compulsory subjects in the British school curriculum.

Perhaps it is the sheer longevity of mathematics that has made it such an integral part of every modern child's education. It is also the case that those who are to progress to the frontiers of the subject need, like young musicians, to start young and practise regularly.

Chapter 5
Mathematical livelihoods

Any mathematician who wants to break new ground needs time to think and scribble, and some kind of financial support. Let us return for a moment to those we met in the first chapter. We have no idea how Diophantus earned his living; perhaps, like many with mathematical talent, he taught. Many of the best-known mathematicians of the century before Fermat also taught mathematics, but often only as a secondary occupation: Girolamo Cardano and Robert Recorde were physicians, though Recorde also worked for much of his life in mints and mines; Rafael Bombelli and Simon Stevin were both employed on practical construction projects; François Viète, like Fermat, was a lawyer and Counselor. Fermat has often been described as an 'amateur' mathematician, but he lived at a time when there were so few professionals that the concept of amateur was meaningless. Wiles, on the other hand, cannot be described as anything but a professional, fully accredited, and paid to work full time in the research and teaching of mathematics.

Over the centuries, there have been significant changes in the ways that mathematicians have been employed. A modern mathematician is very likely to work in education, finance, or industry, all of which are institutionally organized. Some individuals may also be prepared to pay for mathematical services, for tuition, perhaps, or accounting skills, but they keep no more than a small number of people in work. In the 1st millennium AD the picture was very different. Economic and political power throughout most of Europe and Asia was concentrated in the hands of kings, bishops, caliphs, and warlords. Those who wanted to live by their intellectual skills, including mathematics, were wise to place themselves under a patron powerful enough to pay and protect them. Such patronage could take many different forms. In this chapter we will see it at work first in the lives of three scholars of the 10th and 11th centuries in the lands then dominated by Islam.

Patterns of patronage

Thābit ibn Qurra was born in AD 826 in the town of Ḥarrān, very close to the modern Turkish–Syrian border, and spent his early life there as a money changer. He was not a Muslim but belonged to a local sect, the Sabians. Only a few years previously, the library known as the Bayt al-Ḥikma (House of Wisdom) had been established in Baghdad by the Abbasid caliph al-Ma'mūn for the purpose of preserving, and translating into Arabic, texts in Greek, Sanskrit, or Persian. Ibn Qurra's knowledge of Greek and Arabic besides his native language of Syriac brought him to the attention of the Baghdad mathematician Muḥammad ibn Mūsā when the latter passed through Ḥarrān on his return from Byzantium. Unfortunately, we do not know the date of this meeting, but we may suppose that ibn Qurra was still relatively young because at ibn Mūsā's invitation he moved to Baghdad, where he was educated by ibn Mūsā and his two brothers (collectively known as the Banū Mūsā) in mathematics and astronomy.

In the years that followed, ibn Qurra became one of the most respected scholars in Baghdad. He wrote on medicine, philosophy, and religion, but is best remembered now for his work in mathematics and astronomy. He translated several treatises of Archimedes into Arabic, and also wrote extensively on topics that had interested Archimedes: mechanics, and problems to do with areas, surfaces, or volumes of curved shapes. He commented on Ptolemy's *Almagest* and wrote on spherical geometry and astronomy, especially on the motion and apparent height of the Sun, and the motion of the Moon and the five then known planets. He also studied Euclid's *Elements* intensively; his attempted proof of one of Euclid's postulates, on parallel lines, was taken up again in Oxford in the 17th century. Ibn Qurra also produced his own proofs of Pythagoras' Theorem, one of which is shown in Figure 8.

Ibn Qurra remained in Baghdad until his death in AD 901. He retained his association with the Banū Mūsā for many years and taught ibn Mūsā's sons. During the final ten years of his life, he became a regular attendant at the court of the caliph al-Muʻtaḍid, so intimate with the caliph that, according to a biographical sketch by the 12th-century writer, al-Qiftī, he was allowed to 'sit down in his presence at any time he wished'. Later his son, Sinān, and two grandsons became well known scholars in their own right. In what is known of ibn Qurra's life, we may thus discern two crucial features. One is a network of teaching and learning established between friends and families, in this case linking members of ibn

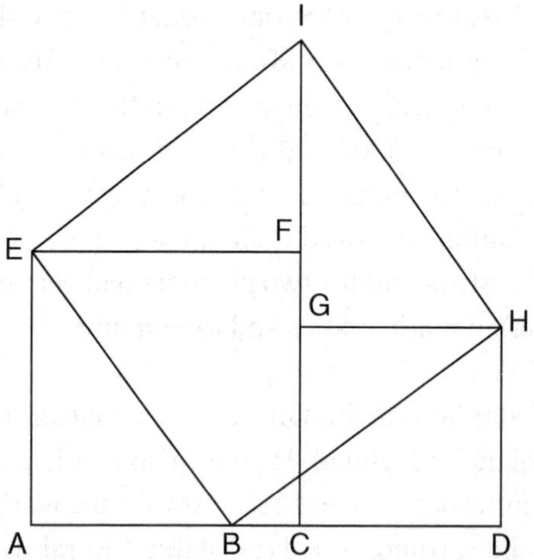

8. Thābit ibn Qurra's proof of Pythagoras' Theorem: a simple cut-and-paste argument shows that IHBE = EFCA + GHDC

Mūsā's family to ibn Qurra's. Such close personal relationships have now been observed several times in the course of this book. The second feature, more particular to ibn Qurra's time and place, is the protection and patronage offered first by the Banū Mūsā, later by the caliph himself.

Another scholar, Abū Rayḥān al-Bīrūnī, generally known as al-Bīrūnī, was born 70 years after ibn Qurra died, at the further end of the Islamic domains, and into a less stable world. The town of his birth on the Āmū Daryā (or River Oxus) lies just inside modern Uzbekistan and is now called Biruni. He was taught by the mathematician and astronomer Abū Naṣr Manṣūr, with whom he continued to work later in his life. As a young man he was already using solar observations to calculate latitudes of local towns, but this activity was disrupted when civil war broke out in 995 and he was forced to flee. We know something of his wide-ranging movements over the next 30 years from his precise observations of eclipses. At times, he worked in the region south of the Caspian Sea, close to modern Tehran, where he is known to have dedicated a text on chronology to the Ziyārid ruler of the region, Qābūs. At other times he lived in his home area, at first under the patronage of the Sāmānid ruler Manṣūr II, later, for 14 years, under that of Abu'l-Abbās Ma'mūn.

This last relatively stable period came to an end in 1017 when the region was overrun by the Ghaznavid dynasty based in what is now eastern Afghanistan. Al-Bīrūnī appears to have been taken prisoner, and subsequently lived for many years near Kabul or at Ghazna itself, about 100km to the south. His relationship to the sultan, Maḥmūd, is unclear: he complained of ill treatment but was later supported in some of his studies. He was also able to travel to northern India, which had also fallen under Ghaznavid rule, and wrote extensively about the region and its religion, customs, and geography. After Maḥmūd's death in 1030, al-Bīrūnī came under a second Ghaznavid patron, Maḥmūd's son Ma'sūd;

and under a third, Ma'sūd's son Mawdūd, after Ma'sūd was murdered in 1040. Al-Bīrūnī himself died in Ghazna in 1050.

Throughout a life beset by dynastic changes, al-Bīrūnī was a committed scholar and a prolific writer. Around half of his works were on astronomy and astrology, with other texts on mathematics, geography, medicine, history, and literature. Unfortunately, only a fraction of what he wrote has survived.

The third mathematical scholar we will look at is 'Umar ibn Ibrāhīm al-Nīsābūrī al-Khayyāmī, better known in the west as Omar Khayyam. He was born shortly before al-Bīrūnī died, in Nishapur in north-east Iran. His name suggests that he came from a family of tentmakers. By this time, the Iranian region had come under the rule of the Seljuqs, a dynasty of Turkic origin. As a young man al-Khayyāmī travelled east to Samarkand where he wrote an important treatise on equations, dedicated to the chief justice, Abū Ṭāhir. He later spent many years in Isfahan where, under the patronage of the sultan Malik-shāh and his vizier Niẓām al-Mulk, he supervised the observatory and the compilation of astronomical tables. During this same period, like ibn Qurra before him, he also wrote commentaries on Euclid. Unfortunately, the observatory was closed in 1092 after Niẓām al-Mulk was murdered and Malik-shāh died. Eventually, after further changes of ruler, al-Khayyāmī left Isfahan. After spending some time in Merv, about halfway between Isfahan and Samarkand, he finally returned to Nishapur, where he died in 1131.

I cannot resist including one of his *Rubā'iyyāt* (quatrains), not from the Victorian translation by Edward FitzGerald but from a 1998 translation by Shahriar Shahriari.

> The secrets eternal neither you know nor I
> And answers to the riddle neither you know nor I
> Behind the veil there is much talk about us, why
> When the veil falls, neither you remain nor I.

These three simplified case studies do not begin to tell all that could be said about mathematical practice under the medieval Islamic dynasties, but at least a few general points emerge. One is that, just as a few centuries earlier, Greek mathematical writers were to be found anywhere around the eastern Mediterranean but rarely in Greece itself, so those who wrote mathematics in Arabic were scattered across an even vaster region, from modern Turkey to modern Afghanistan, though not Arabia itself. For this reason, historians prefer to call such writers 'Islamic' rather than 'Arabic' but, as the example of ibn Qurra shows, not all were Muslims, nor were their mathematical writings related to their religious views. Nevertheless, they all lived in societies where the practices and culture of Islam were dominant and so the label is probably better than any other.

A second theme that comes through is the precariousness of scholarship in a world of rapidly changing rulers and dynasties. For a boy or young man of talent to have his mathematical skills recognized and nurtured was already a matter of chance and circumstance, as for both ibn Qurra and al-Bīrūnī. His ability to study or travel thereafter might depend very largely on the favour and financial support of a ruler whose own future might be far from secure. Al-Bīrūnī seems to have been particularly remarkable in enjoying or surviving the attentions of patrons from opposing dynasties. Despite such difficulties, the output of some of these scholars was both prolific and varied. Those who wrote on astronomy and astrology might also write on spherical geometry and trigonometry, or on the *Elements* of Euclid or the works of other Greek writers, or on arithmetic and algebra, or on geography, history, music, philosophy, religion, or literature.

Finally, one might ask what was in such arrangements for the patron? Individual cases differed greatly, indeed there was no single word in Islamic societies for the relationship here described as 'patronage'. As we have already seen in China and Europe, rulers often valued the mathematically adept for their ability to calculate auspicious dates. In some cases, they may also have hoped for long-term, even eternal, benefit from their support for good works. Moreover, owning the services and companionship of the intellectually talented would have been both a source of pleasure and a sign of prestige.

From about the end of the 12th century, scholars were more frequently able to obtain paid positions at endowed teaching institutions, the *madrasas*, and so became less dependent on the whims or preferences of individual rulers. To examine more closely the shift from patronage to professional employment, however, we will now turn to England at a slightly later date.

From patronage to professionalism

In England, the 40 years from 1580 to 1620 were a period of transition, when patronage still existed but where we may also discern the first signs of movement towards publicly accountable paid positions. The careers of Thomas Harriot, William Oughtred, and Henry Briggs illustrate some of the possibilities and opportunities open to the mathematically talented in England at that time.

Thomas Harriot, born in 1560, studied at Oxford in the years 1577 to perhaps 1580. He did not take a degree in mathematics (there was no such thing at the time) but may have learned something of the subject from tutors or his own reading. Better attested is his interest in exploration and navigation, which he seems also to have acquired at Oxford, possibly from the lectures of the adventurer Richard Hakluyt. During the 1580s Harriot came under the patronage of Walter Raleigh, who at the time was greatly interested in the potential colonization of America. In 1585 Harriot sailed to the coast of what is now North Carolina on a voyage financed by Raleigh, a year-long expedition that ended in failure but which enabled Harriot and his friend John White to bring back a great deal of useful information and some beautiful drawings of the people, flora, and fauna of the region. Unfortunately, he also brought back a fondness for tobacco, which was eventually to kill him.

Before the trip, Harriot had been engaged by Raleigh to teach navigation to the sailors, though unfortunately the text he wrote is now lost. On his return, he continued to live under Raleigh's patronage, at first on Raleigh's estates in Ireland (another colonial venture), later at Raleigh's London home, Durham House on the banks of the Thames. It was from the roof of Durham House that Harriot conducted his early experiments on falling bodies, comparing the rates of fall of balls of metal and wax. Harriot continued to be close to Raleigh right up to the day when Raleigh was executed in 1618: notes on Raleigh's final speech from the scaffold survive in Harriot's handwriting amongst his personal and mathematical papers. By the early 1590s, however, Harriot had a second patron in Henry Percy, ninth Earl of Northumberland. Harriot spent the remaining 30 years of his life at Percy's London home, Syon House in Middlesex on the Thames, or at his country home, Petworth House in Sussex. Unfortunately, neither of Harriot's patrons successfully negotiated the political and religious tensions of the day: Percy, like Raleigh, spent many years imprisoned in the Tower of London. Nevertheless, he supplied Harriot with an income and the freedom to pursue whatever studies he chose. Harriot never lost his interest in the problems of navigation at sea; he also later turned to astronomy, and contemporaneously with Galileo used a telescope to observe sunspots and the craters of the Moon. Through one of his Oxford friends, Nathaniel Torporley, he managed to obtain the mathematical works of Viète (which later so profoundly influenced Fermat), so becoming one of the first people anywhere, and certainly the first Englishman, to appreciate and extend some of the exciting new mathematical ideas that were being developed in France.

Harriot published none of his findings. With a secure private income, he had no need either to prove himself or earn a living. Nor did he teach, though he did discuss his ideas within his own circle of friends. In one sense, Harriot's work had little immediate influence; he certainly did not cause the kind of intellectual stir that Galileo later did. On the other hand, his freedom to work as he chose enabled him to explore a wide range of subjects, some of them fairly arcane, and to carry his investigations to some important conclusions. The modern term for this is 'blue skies research'. Harriot's work could easily have been lost, but fortunately his reputation amongst his contemporaries was such that after his death in 1621 his papers were preserved, and some of the ideas in them continued to circulate amongst his successors over many years. In that sense, Harriot can be said to have encouraged, albeit indirectly, both mathematical discussion and the respect for mathematical and scientific studies that characterized the fledgling Royal Society half a century later. Indeed, such was Harriot's reputation that in its first ten years the Royal Society more than once instigated searches for his surviving papers.

Less creative than Harriot but in some ways of equal importance to the later flourishing of mathematics in England was William Oughtred. Born in 1573, Oughtred was only a few years younger than Harriot but outlived him by about 40 years. From 1610 or earlier, Oughtred was a clergyman in Albury in Surrey; he seems never afterwards to have moved away from there apart from occasional visits to London. He became renowned as a teacher of mathematics, to both children and adults. Like Harriot, he also acquired an aristocratic patron, Thomas Howard, Earl of Arundel, whose country seat at West Horsley was only a few miles from Albury. Oughtred taught Howard's son William as he taught the sons of other local gentry. Through Howard, he also met a relative of the family, Sir Charles Cavendish, who proved to be an important figure in English mathematics at this period. Cavendish was not particularly good at mathematics but for some reason was fascinated by it, and avidly collected and tried to understand the latest books and papers. After Harriot's death, for instance, he copied out entire sections from Harriot's manuscripts, though, he admitted, 'I doute I understand not all'. It was Cavendish who brought Oughtred the work of Viète from France just as Torporley had earlier brought it for Harriot.

It was also Cavendish who encouraged Oughtred to write his first textbook, dedicated to his pupil, the 14-year-old William Howard. The book, published early in 1631, became known by its abbreviated title, *Clavis mathematicae* (*The Key to Mathematics*), and it ran and ran, through five Latin editions and two English translations. The content was elementary, an introduction to arithmetic and algebra, but by that time Recorde's earlier textbooks were almost a century old and there was a desperate need for something fresh. When new professors were installed at the University of Oxford after the years of civil war, it so happened that both were pupils or readers of Oughtred and they immediately introduced the *Clavis* to Oxford, making it the first mathematical book to be printed by the university. Almost every 17th-century mathematician of note, and many who were not, took some of their first steps with the *Clavis*, Christopher Wren, Robert Hooke, and Isaac Newton among them. Thus although Oughtred himself never made any great mathematical advances and taught only at a relatively elementary level, he, like Harriot, indirectly encouraged the dissemination and development of mathematical expertise in early modern England.

Neither Harriot nor Oughtred could have done what they did, however, without the support of the three aristocrats who encouraged their work: Henry Percy, Thomas Howard, and Charles Cavendish. A later member of the Cavendish family gave his name to the Cavendish Laboratory in Cambridge, but the Percy and Howard families have not usually been associated with science or mathematics; nevertheless, without the trust and the intellectual and financial support offered by these three men, it might have taken very much longer than it did for a mathematical community of critical size to emerge in England during the first half of the 17th century.

At the same time, and by contrast, we should not overlook certain other contemporary developments. In 1597, a legacy left by the merchant and financier Thomas Gresham paid for seven public lectureships (one for each day of the week) in astronomy, geometry, physic (medicine), law, divinity, rhetoric, and music. Gresham College (which survives to this day and still offers public lectures) played its own role in strengthening the London mathematical community; meetings held after the lectures during the 1650s helped to bring about the establishment of the Royal Society a few years later. Twenty years after the founding of the Gresham lectureships, Henry Savile founded the chairs of geometry and astronomy at Oxford. For many years there was fluid movement between the posts at Gresham and Oxford. In particular, the first Gresham professor of geometry, Henry Briggs, also became the first Savilian professor of geometry at Oxford.

Briggs, from Halifax in Yorkshire, was almost exactly the same age as Harriot and entered St John's College, Cambridge, in 1577, the same year that Harriot was registered at Oxford. Unlike Harriot, however, he followed a university career, lecturing at Cambridge first in medicine, later in mathematics, before he moved to Gresham College in 1597. He was there for over 20 years until he took up the Savilian chair in Oxford, where he remained until his death in 1630.

Briggs and Harriot make a fascinating pair: one of the tantalizing unanswered questions in the history of mathematics of this period is whether they ever met. They ought to have done. During the years before and after 1600, Briggs, like Harriot, was intensely interested in problems of navigation. In 1610, while Harriot was observing sunspots, Briggs was working on eclipses. When John Napier brought out his 'wonderful invention' of logarithms in 1614, both Harriot and Briggs soon became aware of it. Briggs immediately travelled to Scotland to visit Napier, and helped him to develop his work further; Harriot no longer made long journeys and in any case was already becoming seriously ill, but he made notes on logarithms, and almost certainly recognized their relevance to much of his own earlier work.

One cannot help thinking that with Harriot, as with Napier, Briggs could have engaged in lengthy and fruitful conversations. It could so easily have happened because for the last 20 years of Harriot's life they lived not far distant from each other: Harriot at Syon House, Briggs close to Bishopsgate, only a mile from the Tower of London where Harriot regularly visited Raleigh and Percy. There is no evidence, however, that their lives ever coincided. Their circles of friends and spheres of influence were quite different: Briggs was employed by a public institution, while Harriot worked privately from his own home. A treatise by Briggs on 'The northwest passage to the South Sea through the continent of Virginia', which would surely have interested Harriot, was published in 1622, the year after Harriot's death, and Briggs's *Arithmetica logarithmica* not until 1624. During the 1620s, Briggs did come into contact with Harriot's friend Nathaniel Torporley and was aware of attempts to publish some of Harriot's papers, but he himself died in 1630, the year before the posthumous publication of Harriot's *Praxis*. Thus in print, as in life, they sailed close but managed to miss each other.

The lives of Harriot and Briggs offer a pertinent contrast between the older habits of patronage and the new lives of professional mathematicians, properly paid in return for clear responsibilities, particularly in teaching. The latter, of course, was the way of the future.

Institutions, publications, and conferences

The life of Joseph-Louis Lagrange, one of the finest mathematicians of the 18th century, epitomizes some of the new possibilities opening up to a talented mathematician in western Europe 150 years after the deaths of Briggs and Harriot. Lagrange was born in 1736 into a French-Italian family in Turin (his baptismal name was Giuseppe Lodovico Lagrangia). At the age of 17 he discovered a predilection for mathematics and two years later was appointed to teach at Turin's Royal Artillery School. Lagrange was still living with his family in his home town, but intellectually he had already begun to move further afield. Shortly before he took up his teaching post, Lagrange had sent some of his work to Leonhard Euler, director of mathematics at the Royal Academy of Sciences in Berlin. Further letters to Euler rapidly resulted in Lagrange's election to foreign membership of the Academy. At the same time, he and others founded their own scientific society in Turin, one of many such societies founded in western European cities during the 1750s and the forerunner of the present Academy of Sciences of Turin.

The rise of scientific societies and academies is one of the defining features of the intellectual history of the 18th century. The Royal Society of London had been founded in 1660 and the Academy of Sciences of Paris in 1699; a Prussian Academy of Sciences followed in 1700, restructured as the Royal Academy of Sciences of Berlin in 1740, while the St Petersburg Academy of Sciences was founded on the Parisian model in 1724. These institutions offered employment to a small number of mathematicians and scientists; more importantly, their regular meetings provided a forum for the presentation and discussion of new research. Papers presented to such meetings were later published in the academy's *Acta* or *Mémoires*; this process could take some time, but the resulting volumes eventually circulated to readers throughout Europe, and numerous important exchanges were carried out through the pages of academy journals. Lagrange published most of his own early research in the *Mélanges de Turin*, the journal of his own society in Turin.

The Paris Academy also established a tradition of prize questions, with a period of two years for the response. Lagrange sent in entries for the prizes of 1764 (on why the Moon always shows the same face) and 1765 (which he won, on the movement of the satellites of Jupiter). By this time, therefore, he was becoming known and respected by the leading mathematicians of Europe. Jean le Rond d'Alembert, for example, who had earlier been the scientific editor of the *Encyclopédie*, tried hard to find him a post beyond Turin. In 1766, Euler left Berlin for the Academy of St Petersburg and offered to secure a place for Lagrange in Russia, but Lagrange settled instead for Euler's old position at the Academy of Berlin.

The long relationship between Euler and Lagrange, begun before Lagrange was 20, thus remained both intimate and distant. Euler, the most prolific mathematician of the 18th century, threw out one brilliantly intuitive idea after another, but did not always linger long enough to work them through before moving on to the next thing that caught his imagination. The person who very often followed in his wake, turning half-worked ideas into sound and beautiful theories, was Lagrange. Nevertheless, the two never actually met; indeed, Lagrange always retained a respectful distance from Euler, whom he regarded as his elder and superior. He refused to compete directly with Euler for the Paris prize of 1768 (on the motion of the Moon), though they eventually shared the prize of 1772 on a similar subject. Lagrange remained in Berlin for 20 years, during which time he published extensively (in French) in the Academy's *Mémoires*.

After the death of Frederick the Great, who had done so much to support the Berlin Academy, Lagrange moved once again, this time to the Paris Academy, where he arrived in 1787. Two years later, every institution in France was thrown into turmoil by the Revolution, but Lagrange somehow managed through those years to keep his head and his reputation. In 1795 the Academy was abolished and replaced by the Institut National; Lagrange was elected chairman of the section dealing with physical and mathematical sciences. At the same time, the Revolution's pressing need for properly trained teachers and engineers led to the founding of new institutions, in particular the École Polytechnique in 1794 and the École Normale for the training of teachers in 1795. Lagrange taught at both. The École Polytechnique was to become the most prestigious educational institution of early 19th-century Paris. Anyone who has studied mathematics beyond school level will almost certainly be familiar with the names of Lagrange, Laplace, Legendre, Lacroix, Fourier, Ampère, Poisson, and Cauchy, all of whom taught or examined at the École Polytechnique in its early years. Further, the École published its lecture notes in 'cahiers' which were used as textbooks throughout France, especially by those aspiring to be accepted as students.

Lagrange died in 1813. In the first two-thirds of his career, in Turin and Berlin, he had both contributed to and benefited from national academies and their respective journals, institutions which had done much to foster the creation and dissemination of new research. During his final years in Paris, Lagrange saw the rise of a new kind of institution, designed to offer a high level of mathematical and scientific training to the most able students. Unlike the universities, the École Polytechnique offered an education that was tightly focused and practical, and would enable its graduates to consolidate the gains of the Revolution and later the Napoleonic empire.

In case a history of institutions seems somewhat impersonal, let us not lose sight of the close personal relationships that also ran through Lagrange's life, notably with Euler and d'Alembert. When Lagrange died, his protégé, Augustin-Louis Cauchy, son of a family friend, was just embarking on his own long career and was to be a leading figure in French mathematics until his death in 1857. It is possible to trace unbroken chains of personal friendships and collaborations in western European mathematics, from Leibniz in the late 17th century through the Bernoulli family and Euler to Lagrange and on to Cauchy in the mid-19th century.

By the time of Lagrange's death, changes were under way in his earlier home, Berlin. The University of Berlin was founded in 1810 by Wilhelm von Humboldt, as an institution that would not simply pass on accumulated knowledge but which would encourage and facilitate new research. German university professors were free to make their own appointments and so to determine the direction and emphasis of their departments. Research groups, seminars, and doctoral training were all established in the German universities before 1900 and are now imitated, more or less, in every university in the world. Academic mathematicians, Andrew Wiles among them, are all in that sense a product of 19th-century Germany.

The publication of mathematical research changed too. In the 17th and 18th centuries, the main outlets for mathematical articles had been academy journals. The first printed mathematical paper appeared in the *Philosophical Transactions of the Royal Society* in 1668, written by the then president of the Society, William Brouncker. That paper was only four pages long and was juxtaposed with letters to the editor on 'Chymical, Medicinal and Anatomical particulars', on 'the Variety of the Annual High-Tides', and some miscellaneous notices of new books. Journals later became rather better organized: the *Acta eruditorum*, for instance, had separate sections on medicine, mathematics, natural philosophy, law, history, geography, and theology; but scientific journals throughout the 18th century continued to publish on a wide range of subjects of which mathematics was just one.

The first journal dedicated to mathematics alone, the *Annales de mathématiques pures et appliquées*, was founded and edited by

Joseph Gergonne in France in 1810, and became known as Gergonne's journal. Note here the first appearance of a distinction that had not up to then existed in any formal sense between 'pure' and 'applied' mathematics. Gergonne's journal lasted only until 1832, but by then its German equivalent, with a parallel title, had been established in 1826 by August Crelle. The *Journal für die reine und angewandte Mathematik* (Crelle's journal) exists to this day. So does the replacement of Gergonne's journal, first edited by Joseph Liouville in 1836, the *Journal de mathématiques pures et appliquées* (Liouville's journal). Mathematical journal publishing has continued to flourish and increase ever since: today, journals no longer specialize in mathematics as a whole but in the branches and twigs of the discipline. One of the titles I like is the *Journal of Ill-Posed and Inverse Problems*, but there are hundreds of others.

Specialized institutions, entrance examinations, lengthy training, dedicated journals, professional societies, and regular meetings and conferences, are the hallmarks of every modern profession, including mathematics. International or even national conferences did not exist in Lagrange's day, but they certainly do now, and absorb at least some of the time of all academic mathematicians. In particular, mathematicians are always ready to celebrate each other's important birthdays, another sign of the strong social cohesion of the discipline.

The first International Congress of Mathematicians was held in Zürich in 1897, and was attended by representatives of several European countries and the United States. The second Congress, held in Paris in 1900 to coincide with the Exposition Universelle, is best remembered for a speech made by the German mathematician David Hilbert, in which he outlined 23 problems that he hoped mathematicians would solve in the new century (though proving Fermat's Last Theorem was not one of them). After 1900, Congresses were held every four years except during World Wars I and II. The exclusion of mathematicians from Germany, Austria, Hungary, Turkey, and Bulgaria during the

1920s, however, and the absence of others who objected to that ruling, led to debate as to whether these congresses could be called 'international'.

A list of cities that have hosted the Congress tells its own story of the increasingly global nature of mathematical research. Until the 1960s, all the meetings were held in western Europe, Canada, or the United States, but the Congress for 1966 was held in Moscow and for 1982 in Warsaw. The first Asian country to host it was Japan, in 1990, followed by China in 2002 and India in 2010. When Wiles announced his proof of Fermat's Last Theorem in his home city of Cambridge, he could just as easily have spoken to similar audiences in Beijing, Madrid, or Hyderabad, the venues of the three most recent Congresses. Mathematics is now not only a highly professionalized discipline but a thoroughly international one.

By now we have reached the top of the mathematical pyramid, the tightly knit community of professionals that has become associated with the words 'mathematics' and 'mathematician'. Compared to the number of people from schoolchildren upwards who regularly practise mathematics, however, this professional community is tiny, and the number of women in it is even tinier. One can wonder why women are still so under-represented. There is no easy answer to this question, but we should recall that as in most professional spheres the rules were devised by and for men, and it may be that some women find the air at the top of the pyramid a little rarefied and the company not always congenial. If we leave elite mathematics to elite historians, this hardly matters. Just as mathematics itself has gone through many manifestations, so mathematical lives have been lived in a multitude of ways, none more valid or correct than any other.

Chapter 6
Getting inside mathematics

So far, I have avoided too much discussion of mathematical technicalities, and we will not go far into them in this chapter either, but a historian of mathematics is bound to engage not only with the social context of mathematical texts written in the past but also as closely as possible with their content. This is more easily said than done. At one level, the mathematics of the past can seem easy compared with what is expected of, say, a college student today. The difficulty for the historian is usually not so much understanding the mathematics itself as entering into the mind and mathematical universe of someone from another era.

As an example, let us think for a moment about Pythagoras' Theorem, which has now been mentioned several times in this book. Euclid's proof of the theorem is illustrated in Figure 9. It entails drawing the squares on the three sides of the right-angled triangle, dividing the largest one into two parts, and then showing that each of those parts is equal to one of the two smaller squares. The details were cleverly shown in colour by Oliver Byrne in 1847, in the almost wordless proof shown in Figure 10. One of the key features of this proof is that it applies to any right-angled triangle, however you happen to draw it (indeed, David Joyce's interactive version will allow you to push and pull the original triangle around as much as you like as long as you keep the right angle). In other words, the proof does not depend on particular measurements;

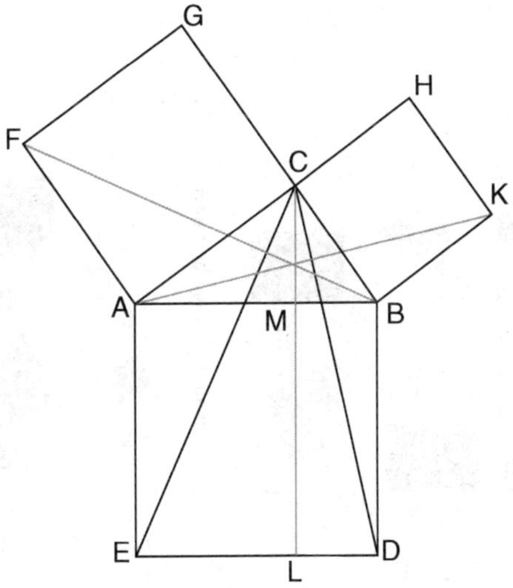

9. Euclid's proof of Pythagoras' Theorem: AFGC = AMLE and CHKB = BDLM

there is no arithmetic involved, and certainly no algebra. This is completely in keeping with the style of the *Elements*: Euclid allowed his readers a straight edge and a compass but no calculator.

Thābit ibn Qurra's proof, shown in Figure 8, also relies on cut-and-paste geometry to demonstrate that the larger square can be made to cover the two smaller ones. For Euclid and ibn Qurra, the underlying intuition behind both the theorem and its proof was geometric.

Now consider the modern technique of labelling the sides of the triangle a, b, and c and writing down $a^2 = b^2 + c^2$. Does this represent the theorem that Euclid had in mind? In one sense, yes. We know that the area of a square with side a is a^2, so the formula is just a very concise way of encapsulating a geometric fact. There is even continuity in the language: we use the same word 'square' for the quantity a^2 and for the four-sided geometric shape. But in another sense, no. The formula comes from a mathematical

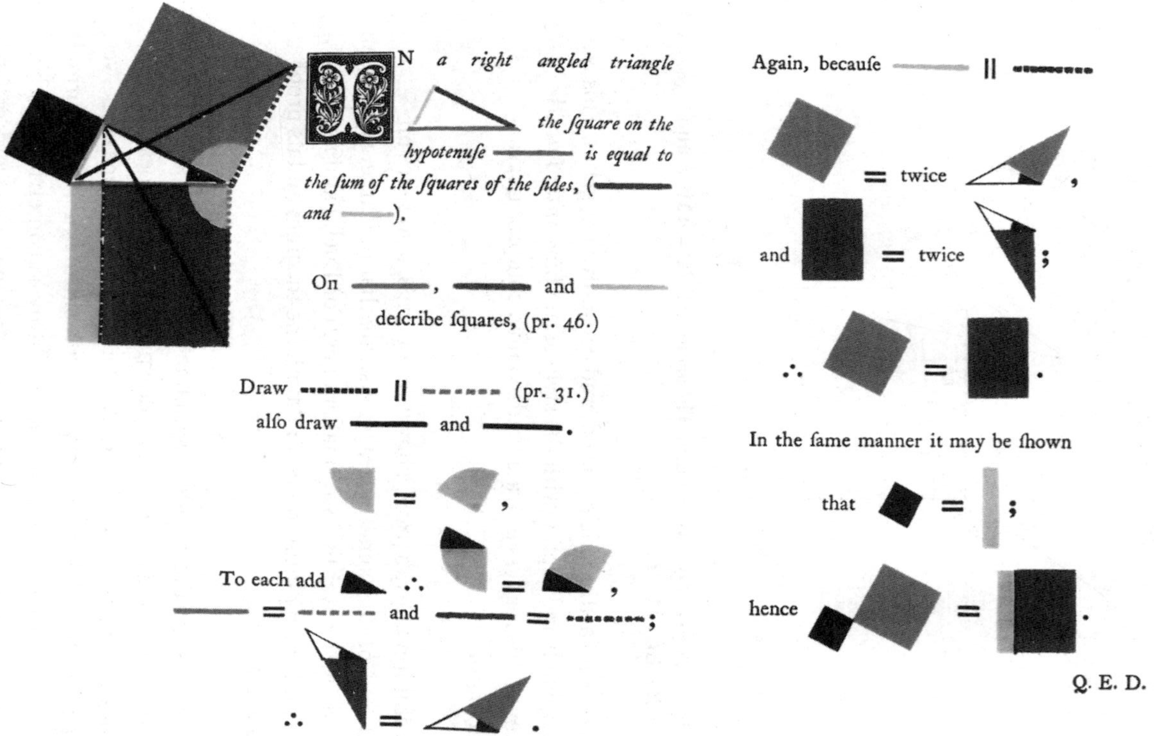

10. Oliver Byrne's proof of Pythagoras' Theorem

culture very different from Euclid's, in which we have learned to let letters represent lengths, and in which we can even forget about geometry and manipulate the letters according to their own rules. Thus, if we wish, we can re-write the above formula as $c^2 = a^2 - b^2 = (a - b)(a + b)$, which is true but no longer has any obvious relevance to a right-angled triangle.

The change from geometric insight to algebraic manipulation is not a trivial one: it takes some effort to learn how to do it. Historically, the shift from a mathematical culture in which geometry was dominant to one in which the language of algebra began to take precedence came about in western Europe in the 17th century. (Fermat was one of the earliest mathematicians to experiment with the possibility, though later he also complained bitterly about any departure from traditional ways of doing things.) Historians have studied this period intensively because the changes were crucial to the development of modern mathematics. To regard the algebraic version of Pythagoras' Theorem as essentially the same as the geometric version is to ignore the historical gulf between them, a gulf that has been crossed only by the cumulative endeavours of many individual thinkers.

Reinterpretation

The example we just looked at is a case of mathematical reinterpretation, in this instance taking a geometric theorem and reinterpreting it algebraically. This is something that mathematicians do a lot; indeed taking an earlier piece of work, their own or someone else's, exploring it, extending it, trying it out under new conditions, is one of the chief ways that mathematicians develop their subject. It is one thing for mathematicians themselves to re-write old mathematics, however, and quite another for historians to do it. When the *Arithmetica* of Diophantus was rediscovered in Europe during the

Renaissance, it turned out to be such a rich source of problems that it was reinterpreted in several ways, both mathematically and historically. We will look at some of the mathematics first.

We have already seen how Fermat extended Problem II.8, testing it out on cubes or even higher powers as well as squares. Here we will look at another early 17th-century reinterpretation, of a different problem from Diophantus, this time by the English mathematician John Pell. Born in Southwick in Sussex in 1611, Pell lived at the same time as both Harriot and Oughtred (whom we met in Chapter 5), though he was some 50 years younger. He was educated at the newly founded Steyning Grammar School, a few miles north of Southwick, and then at Trinity College, Cambridge. Afterwards he returned to Sussex and taught at an experimental school in Chichester until it closed a few years later. Pell then spent several years searching for either a paid post or a patron but found neither that suited his rather particular temperament. At the end of 1643 he was appointed to the Gymnasium in Amsterdam, and two years later to the Illustre School in Breda where he remained until 1652.

During this time Pell gave a good deal of attention to Diophantus. We know this because by the early 1640s Pell had become acquainted with Sir Charles Cavendish (whom we also met in Chapter 5) and they corresponded throughout Pell's years in the Netherlands. They had their own version of a mathematical patronage relationship: Cavendish would ask Pell to help him with whatever he had failed to understand in his latest effort at mathematical reading and Pell would duly respond. Cavendish clearly thought highly of Pell's abilities and expected him to publish a number of important books, including a new edition of Diophantus 'which', he wrote, 'I am exceeding greedie to see'. Unfortunately, Pell was almost pathologically incapable of finishing or publishing anything, but there is evidence that he at least began to work on such an edition.

That evidence comes from Pell's voluminous notes (thousands of pages now bound in more than 30 large volumes in the British Library). The reason that Pell was so interested in Diophantus was that he, Pell, had developed a method of problem-solving that he thought perfectly suited to the questions in the *Arithmetica*. The method was this: first, for any question set down the unknown quantities and the given conditions in numbered lines; second, work systematically from the conditions to the required answer. To ensure that the work proceeds properly, it is set out in three columns, with line numbers in the narrow central column. For each line, the left-hand column contains a brief instruction; the right-hand column shows the result of carrying it out. The whole thing has much the look and feel of a modern computer algorithm.

To see how such a method can be applied to the ancient work of Diophantus, let us look at Pell's rendition of Problem IV.1 of the *Arithmetica*: to find two numbers whose sum is a given number and whose cubes sum to another given number. Diophantus suggested that the sum of the two numbers should be 10 and the sum of their cubes 370, precisely the problem later worked by the young Anne Davenant under her father's instruction. Pell solved it in his own unique style. His first two lines are the following, in which he calls the unknown numbers a and b.

$a = ?$	1	$aaa + bbb = 370$
$b = ?$	2	$a + b = 10$

Next, following Diophantus exactly, Pell introduced a third number, c, and supposed that $a = c + 5$ so that, necessarily, $b = 5 - c$. His next two lines are therefore

$c = ?$	3	let $a = c + 5$
$2' - 3'$	4	$b = 5 - c$

where $2' - 3'$ just means subtract line 3 from line 2. Now everything is set up and the work can proceed. The reader who wants to follow the detail needs to know that Pell's instruction $3'$@3 means take the cube of line 3, while $10'\omega 2$ means take the square root of line 10. A further convention favoured by Pell is switching from lower- to upper-case letters once the required values have been found.

$3'@3$	5	$aaa = ccc + 15cc + 75c + 125$
$4'@3$	6	$bbb = 125 - 75c + 15cc - ccc$
$5' + 6'$	7	$aaa + bbb = 30cc + 250$
$7', 1'$	8	$30cc + 250 = 370$
$8' - 250$	9	$30cc = 120$
$9' \div 30$	10	$cc = 4$
$10'\omega 2$	11	$c = 2$
$11' + 5$	12	$c + 5 = 7$
$5 - 11'$	13	$5 - c = 3$
$3', 12'$	14	$A = 7$
$4', 13'$	15	$B = 3$

The final four lines check that the problem has indeed been correctly solved:

$14'@3$	16	$AAA = 343$
$15'@3$	17	$BBB = 27$
$16' + 17'$	18	$AAA + BBB = 370$
$14' + 15'$	19	$A + B = 10$

It seems that Pell planned to re-write the whole six books of the *Arithmetica* in this style, but if he ever completed them his manuscript is lost. Many of his contemporaries were impressed with his method, though. His friend John Aubrey even invented a new Latin verb for it: *pelliare*, 'to pelliate'.

It is clear from the above example that Pell did not believe in wasting words: the only one that appears in his 19 lines of working is 'let' (he actually wrote it in Latin as *sit*). But if words are to disappear, there must be symbols to replace them, and here Pell was a master of invention. The symbols @ and ω that helped to keep his left-hand column concise have long since fallen out of use, but his division symbol $\div$ is still with us. Invention of notation was one of Pell's particular talents; in this he was following what was something of an English tradition at the time. By 1557 Robert Recorde had devised the = sign based on two parallel lines 'bicause noe 2 thynges, can be moare equalle'. Around 1600 Thomas Harriot had added the inequality signs $<$ and $>$ and the convention of writing *ab* for *a* multiplied by *b*. In 1631 William Oughtred introduced the $\times$ sign, though he rarely used it; he also argued passionately that notation 'plainly presenteth to the eye the whole course and processe of every operation and argumentation'. This was clearly what Pell thought too, that his method rendered the argument plain to the eye without need for further explanation. His efforts to 'pelliate' Diophantus therefore tell us rather more about the aspirations of early 17th-century English algebraists than they do about Diophantus and his *Arithmetica*.

This is true of reinterpretations of a historical rather than mathematical kind too: that they generally reveal more about the interpreter than the interpreted. Stories that have circulated over the centuries about the origins of algebra, for example, have recorded not just historical fact but contemporary understanding. Algebra first came to non-Islamic regions of western Europe in the late 12th century through translations of al-Khwārizmī's *Al-jabr wa'l muqābalā*, but by the 16th century, this early history had been forgotten, if indeed it had ever been properly known. Nevertheless, the Islamic origins of the subject were recognized, if only from the strange-sounding words 'algebra' and 'almucabala' associated with it. Thus early 16th-century writers ascribed the invention of algebra variously to '*une nome Arabo di grande intelligentia*' ('a certain Arab of great intelligence'), sometimes to one Algeber (actually Jābir ibn Aflaḥ, a 12th-century Spanish Muslim astronomer who had nothing to do with it), or to the vaguely named 'Maumetto di Mose Arabo' (a rendering of Muḥammed ibn Mūsā, an Arab).

In 1462, however, the German scholar Johannes Müller, usually known as Regiomontanus after the Latinized name of his home town, Königsberg, examined a manuscript of Diophantus'

Arithmetica in Venice. Three years later, lecturing in Padua, he described the contents as 'the flower of all arithmetic ... which today is called by the Arabic name of algebra'. His lecture was not published until 1537, but very soon after that other writers began to pick up the same theme: that algebra had been invented by Diophantus and was only later adopted by 'the Arabs'. One can see why such stories were accepted at a time when a Greek pedigree conferred instant respectability and status. The fact that the problems handled by Diophantus were different in both style and content from those found in Islamic texts does not seem to have deterred anyone from thinking that the latter must somehow have been derived from the former.

Even today, when there is much greater appreciation of the mathematics that western Europe inherited from the Islamic world, Diophantus is still sometimes credited as being the founder of algebra. This is a debate that can run and run, but we should try to understand mathematically what is at stake. It is true that Diophantus posed several 'find a number' problems that are easily handled by modern algebraic methods, as Pell's example demonstrates. But he also posed a great many other problems that are 'indeterminate', that is, they have more than one possible answer. In such cases, Diophantus was usually satisfied if, by some special method, he could show just one of those answers. In fact his work is full of ideas, some of them very clever, that work for specific questions, unlike the more general rules of the later Islamic algebra texts. It has also been suggested that Diophantus used an elementary symbolic notation by writing, for example, ς for an unknown number and Δ^{Υ} for its square, but it has now been shown that these abbreviations for the Greek words *arithmos* (number) and *dynamis* (square), respectively, were introduced by 9th-century copyists and cannot be attributed to Diophantus at all. Finally, the mathematics that has been derived from the *Arithmetica* has been absorbed into modern number theory, whereas Islamic *al-jabr* texts gave rise much more directly to the algebra of western Europe. It seems to me that the word 'algebra' should be reserved for the rules and procedures that were described by the participants themselves as '*al-jabr*' or 'algebra', and that we should not impose those words, nor the history they carry with them, on a writer who was working in an earlier and very different tradition.

Who was first . . . ?

The question we have just examined, 'Who invented algebra?', is typical of those sometimes asked of historians of mathematics, who are often expected to be able to say who was first to discover or invent certain ideas. Except in the simplest cases such questions can be extraordinarily difficult to answer. Take, for example, the discovery of calculus. This is the branch of mathematics that can be used for describing and predicting change. It is used today in biology, medicine, economics, ecology, meteorology, and every other science that works with complex interactive systems. It is therefore not unreasonable to want to know 'Who invented calculus?'

The short answer is that two people did, almost simultaneously but independently: Isaac Newton working in Cambridge and Gottfried Wilhelm Leibniz working in Paris. To modern historians, there is no longer any dispute about this because we have the manuscripts of both men and can see exactly when and in what order their ideas were developed. We can also see that they approached the work in very different ways, each devising their own vocabulary and notation (Leibniz spoke of 'differentials' while Newton spoke of 'fluxions'; Leibniz invented the now familiar $\frac{dx}{dt}$ notation whereas Newton used the now less common $\dot{x}$).

For their contemporaries, however, the story was not clear at all. The basic facts are that Newton developed his version of calculus during 1664 and 1665 (before his 23rd birthday) but then did nothing with it. By the early 1670s he had already engaged in an intellectual skirmish with Robert Hooke over his optical findings and was perhaps reluctant to risk another over the calculus. In any case, by that time his interest had shifted to alchemy, which was to preoccupy him for the next decade. In 1673, however, Leibniz, then living in Paris, independently began to work on some of the same problems that had earlier intrigued Newton, and published his first paper on calculus in 1684 followed by others in the 1690s. Newton appears to have taken little notice, probably regarding Leibniz's early work as rather trivial compared with what he himself had been able to achieve. Some of Newton's friends felt differently, however, and around the turn of the century his English supporters began to hint not only that Newton had been first but that Leibniz might actually have stolen the seeds of his ideas from Newton. It did not help Leibniz's case that he had seen some of Newton's papers when he was in London in 1675 and had received letters from Newton in 1676, but what he had learned from them, and how that related to what he had discovered already, no-one but Leibniz really knew.

Both Newton and Leibniz held back from direct confrontation but allowed the battle to be fought out through their henchmen, who were thoroughly belligerent on both sides. Eventually, in 1711, Leibniz appealed to the Royal Society, of which he was a member, to adjudicate in the dispute. Newton, as President of the Society, set up a committee which barely needed to meet because Newton was already busy writing its report. Not surprisingly, it found in Newton's favour. And, also not surprisingly, that was not the end of the matter: the dispute rumbled on until after Leibniz's death in 1716. The dispute explains why in 1809 the English schoolboy George Peat in Cumbria learned a subject called 'fluxions' rather than a subject called 'calculus'.

It is an unedifying story from which no-one comes out well. The point of re-telling it is to emphasize how difficult it was for anyone at the time to get to the bottom of it: no single person was in possession of all the facts; besides, it was difficult to know whether the argument was about the calculus as a whole or about particular aspects of it (Leibniz accused the English of shifting ground on this); and, as can be the way with disputes, several disagreements were dragged in that were never part of the original argument. Another point of the story, however, is that the ultimate evidence for the truth comes not from what people at the time wrote or said, which was almost always partial (in both senses of the word), but from the mathematical manuscripts themselves.

In mathematics it is not at all uncommon, as in the case of the calculus, for two people to come up with similar ideas at more or less the same time. Once the groundwork has been laid, one mathematician can make use of it just as easily as another, and it then becomes very difficult to apportion credit, especially if the two have had some contact with each other. It was for precisely this reason that Wiles shut himself away so carefully during his years of work on Fermat's Last Theorem. In the case of the calculus, there is enough documentary evidence for historians to work out what really happened but this is not always so. Two early 19th-century mathematicians, Bernard Bolzano in Prague and Augustin-Louis Cauchy in Paris, also developed some remarkably similar ideas, Bolzano in 1817, Cauchy in 1821. Did Cauchy 'borrow' from Bolzano or not? Bolzano's work was published in a little-known Bohemian journal which was nevertheless available to Cauchy in Paris. On the other hand, both could have built independently on the earlier work of Lagrange. We might also throw into the assessment circumstantial evidence about Cauchy's way of working, which was very often to pick up good ideas from someone else and develop them at length. In the end, for lack of firm evidence either way, we simply cannot say.

Another problem about saying who was first to make a discovery can be defining what we think the discovery actually consists of. At what precise point in history, for example, can we say we have 'calculus', as opposed to a tangle of related ideas that gradually began to make sense first to Newton and later to Leibniz? It is just as difficult, as we have already seen, to pinpoint where algebra began, or where Pythagoras' Theorem became a formal theorem as opposed to a useful fact known to builders. Almost all new mathematics is built on previous work, and sometimes on a number of contributory ideas. Tracing the antecedents of a particular technique or theorem is one of the tasks of the historian, not in order to say who was first, but to understand more clearly how mathematical ideas have changed over time.

Getting things right

Euclid's systematic deductive style, in which every theorem is carefully proved from the theorems and definitions that have gone before it, has stood for centuries as the gold standard of mathematical style. But even Euclid turned out not to be infallible. Questions were raised about one of Euclid's postulates as early as the 5th century AD and proved very difficult to answer. The troublesome postulate is sometimes known as the Parallel Postulate; it can be expressed in different ways, but the simplest is to say that if we have a line *l* in a plane, and a point *P* not on the line, there is just one line through *P* parallel to *l*. Most of us would have no difficulty accepting that. It leads to the consequence that the angles in a triangle add up to 180 degrees, and most of us have no difficulty accepting that either. Many commentators on Euclid, however, thought that the Parallel Postulate should not be a postulate but a theorem, that is, that it should somehow be possible to prove it from the other definitions and postulates. Thābit ibn Qurra and Umar al-Khayyāmī were amongst those who tried; so was John Wallis in Oxford in 1663. Then in 1733 an otherwise little remembered mathematician called Gerolamo Saccheri, professor of mathematics at Pavia in northern Italy, tried a different approach. He investigated what would happen if he assumed that the angles of a triangle add up to either less or more than 180 degrees, hoping, of course, that the results would be absurd so that such assumptions could be dismissed. He was wrong. Assuming that the angles add up to less than 180 degrees led him to some strange but nevertheless consistent results.

A hundred years later, Nicolai Ivanovich Lobachevskii, professor at the university of Kazan in Russia, and János Bolyai from the town that is now Cluj in northern Romania, took these ideas much further (another example of independent but more or less simultaneous discovery); they both realized that it is possible to construct a kind of geometry that is mathematically acceptable but definitely non-Euclidean. The idea was startling to 19th-century thinkers: one consequence was that no-one could know whether infinite space itself was Euclidean or non-Euclidean, any more than we can tell from walking down the street whether the Earth is round or flat. Mathematics was supposed to offer indisputable truths about the world, but suddenly such truths had come to seem less secure.

One of the outcomes of all this was that mathematicians began to look more carefully at their underlying assumptions, formally known as axioms. Indeed, in the late 19th and early 20th centuries, in a return to true Euclidean style, entire branches of mathematics came to be set up on axiomatic foundations, imposing on them a logical rigour that mathematics had not known since the Greek era. For between the 2nd century BC and the 19th century AD, mathematics had developed for the most part in a thoroughly haphazard way. The fact is that mathematicians do not make discoveries by setting up axioms and thinking logically about them but by responding imaginatively to problems that interest them, by opening up questions in new directions, or by seeing how apparently different bits of mathematics might fit together in a fresh way. Of course, they must apply their skills and experience correctly, and in the end they must present a watertight argument known as a 'proof', as Wiles did in his Cambridge lectures, but that is likely to be some way down the road from the initial insights and the hard work that almost invariably follows them.

The discovery of the calculus, discussed in the previous section, is a supreme example of mathematics that in its inception was not logical at all. The whole idea was based on what 17th-century mathematicians called 'infinitely small quantities'. The question one is forced to ask about an infinitely small quantity, however, is: does it have any size at all? If it does, then it cannot be 'infinitely small'; but if it has no size, then it does not even exist and one cannot use it in any sensible way in one's calculations. This may seem like nit-picking, more like a discussion of angels on a pinhead than of mathematics. But it matters, because discussions of infinitely small quantities can rapidly lead to contradictions; and since mathematics is supposed to be a unified logical edifice, a single contradiction brings the whole lot tumbling down. (For this reason mathematicians often deliberately set up a contradiction, as Saccheri tried to do, if they want to prove that something is impossible; the technique is called *reductio ad absurdum.*)

Both Newton and Leibniz were well aware of the paradox of the infinitely small and did their best to deal with it, Newton by addressing it head on, Leibniz by skirting round it. Those who came after them were aware of it too, not only mathematicians but well-educated members of the public also. Bishop George Berkeley, for instance, in a book called *The Analyst: A Discourse Addressed to an Infidel Mathematician* asked 'whether mathematicians, who are so delicate in religious points, are strictly scrupulous in their own science? Whether they do not submit to authority, take things upon trust, and believe points inconceivable?' Did such concerns stop mathematicians in their tracks? No, because very early in the evolution of the calculus they had recognized how powerful it could be and were busy applying it, with plenty of success, to rays of light, hanging chains, falling bodies, vibrating strings, and many other phenomena of the physical world. They were hardly going to give up all this for what they regarded as a metaphysical rather than mathematical difficulty. It took about 150 years until the problem was resolved to most people's satisfaction, in ways that are a little too technical to discuss here; during those same 150 years, however, mathematics advanced beyond all expectations, despite the shakiness in its foundations.

A similar story can be told for the 19th century. In 1822, Joseph Fourier, a lecturer at the École Polytechnique in Paris, published a treatise on the diffusion of heat, his *Théorie analytique de la chaleur*. In it, Fourier investigated the idea of using infinite sums of sines and cosines to describe periodic distributions; these infinite sums are now known as Fourier series, and have a vast range of applications in engineering and physics. Fourier's original derivation, however, was riddled with errors and inconsistencies. Some of these cancelled each other out, but many of them Fourier ignored, if indeed he ever noticed them. In other words, the initial theory of Fourier series was no more firmly grounded than the calculus had been, and yet, like the calculus, it proved to be an immensely rich and useful tool. But just as with the calculus, many mathematicians after Fourier had to spend a great deal of time mending the holes.

These examples are not exceptional. As we have seen, Wiles, a far more competent mathematician than Fourier, had to go through a very similar process of fixing an error, though in his case it took only two years and not a century. Almost every new discovery in mathematics starts out in a rough and ready state and has to be improved and refined before it can be presented to one's peers, let alone taught to beginners.

Most modern textbooks follow the same pattern as Euclid's *Elements*, starting from simple beginnings and building up the mathematics in a seamless logical flow. In other words, we allow – or expect – students to follow a clear-cut path that the first explorers could not even see. If students are given the opportunity to go back to the original discoveries, they are likely to find something very different: a process of trial and error, false starts, and dead ends; half-formed ideas, half worked out, left to be developed by someone else; better-formed ideas refined over months or years; all of it finally adapted by teachers who may not be innovators but who must have the equally important gift of seeing how to explain things to novices. In other words, the polished exposition of a textbook tells us very little about the intuition and hard work, the imagination and struggle, that went into the mathematics in the first place. That is the task of the historian.

Chapter 7
The evolving historiography of mathematics

Ways of doing and thinking about the history of mathematics have changed a great deal over the last few centuries, some of them in keeping with changes in intellectual history more generally, some of them peculiar to mathematics. As we saw in Chapter 2, the approach taken by John Leland in the 1550s, followed by Johann Gerard Vossius a century later, was to record as many facts as possible about authors, dates, and texts, but without any analysis of what those texts contained. By the late 17th century, however, it was clear to anyone interested in mathematics that the power, scope, and techniques of the subject were advancing rapidly: 'Geometry is improving daily', wrote Joseph Glanville in 1668, while just a few years later John Wallis extolled the 'progress and advancement' that had brought algebra to 'the heighth at which now it is'.

The 18th century, the age of the *Encyclopédie*, saw two substantial publications concerned with the history of mathematics, Jean-Étienne Montucla's *Histoire des mathématiques* published in Paris in 1758 (expanded to four volumes in 1799–1802), and Charles Hutton's *Mathematical and Philosophical Dictionary*, which included a number of historical articles, published in London in 1795. By the late 19th century, however, the focus was shifting, as in other areas of study, away from second-hand accounts to scholarly editions and translations of ancient and medieval texts (as had also happened during the Renaissance). To take examples from texts that have been discussed earlier in this book: the first English rendering of the *Arithmetica* of Diophantus was published by Thomas Heath in 1885; Heath's edition of Euclid's *Elements*, based on the best scholarship then available, appeared in 1908; Charles Louis Karpinski's translation of al-Khwārizmī's *Al-jabr* from a medieval Latin version appeared just a few years later, in 1915. Such editions were and remain invaluable: neither the *Arithmetica* nor the *Al-jabr* had previously been available in English; Heath's *Elements* remains the standard English edition to this day.

Modern historians, however, also treat such editions with some care. Heath's essay on the *Arithmetica* is entitled *Diophantus of Alexandria: A Study in the History of Greek Algebra*, a title that raises questions that have already been addressed in this book. Further, a commentator on Heath's edition of Apollonius has observed that 'thanks to skilful compression and the substitution of modern notation for literary proofs, [it] occupied less than half the space of the original'. Again, the historian might not thank Heath for this skill, preferring to see the text uncompressed and free of the anachronisms of modern notation. A good deal of early 20th-century scholarship in the history of mathematics, however, often carried out by mathematicians rather than historians, proceeded in much the same way, translating texts originally written in Egyptian hieroglyphics or in Sumerian, Sanskrit, or Greek into the symbols and concepts of modern mathematics. The motives of the translators were not in themselves reprehensible: in trying to understand ideas that at first seem thoroughly alien, it is natural to try to relate them to something more familiar; the danger is that one comes to regard unfamiliar ideas as no more than archaic renderings of what we can now do, as we see it, more efficiently. In this way, history comes to be re-written from our own perspectives instead of those of the original authors.

Historians of ancient mathematics were among the first to rebel against the distortions introduced by modernization, and during the 1990s led the way in attempting to recover and preserve, as far as possible, the idiom and thought processes of the originals. As Reviel Netz, editor and translator of Archimedes, has said in a remark now often quoted: 'the purpose of a scholarly translation as I understand it is to remove all barriers having to do with the foreign language itself, leaving all other barriers intact'. This forces the modern reader of historical mathematical texts to work much harder than the reader of 50 years ago, but the gains in historical understanding are incomparably greater.

Those who work with ancient mathematical writings have led the way in other aspects of historiography too, partly because of the way their material has been so haphazardly accumulated in the past. A single clay tablet, for instance, tells us very little unless we know where or when it was written. Such information is essential if we are to build up a picture of how a particular text relates to others found in the same area or elsewhere. Many tablets from early excavations were deposited in museums with minimal information on their provenance, or sold on the antiquities market with none, making it exceedingly difficult for historians to deduce useful information from them now. Fortunately, present-day archaeologists record positions and surroundings with great care before removing each layer of evidence. Modern technology has also enabled advances in reading faded or damaged pen and ink texts. Work on the rediscovered text by Archimedes mentioned in Chapter 3 has been particularly remarkable in this respect. Scholars have been able not only to read much of the original text but to identify the scribe who scrubbed the parchment clean and overwrote it: Ioannes Myronas, working in Constantinople during Lent 1229. It is entirely appropriate that recovery of the text and recovery of the story of the text should go hand in hand.

Historians of mathematics have increasingly moved away from a purely 'internalist' view in which mathematical developments are seen to come about of their own accord, regardless of outside influences. As has now been shown over and over again in this book, mathematical activity has for centuries manifested itself in a variety of ways, all of them socially and culturally determined. We should not throw out the baby with the bathwater, however: mathematicians often devote themselves to a particular problem not because it might be useful or because anyone requires them to do so, but because the problem itself catches their imagination. This was precisely the case for Newton and Leibniz with the calculus, Bolyai and Lobachevskii with non-Euclidean geometry, or Wiles with Fermat's Last Theorem. In such cases, progress depends first and foremost on deep and concentrated engagement with the mathematics, and in that sense mathematical creativity can be said to be an internal process. But the mathematical questions that are considered important at a particular time or place, the way they have come to be there, the way they are understood and interpreted, are all influenced by a multitude of factors outside the mathematics itself: social, political, economic, and cultural. Context has become as important to the historian as content.

Another significant change in recent years has been the growing recognition that the mathematics done by a small number of famous mathematicians has not reflected (though it has built on) the diversity of mathematical activity and experience at other levels of society. The history of non-elite mathematics has been one of the key themes of this book. Historians of mathematics, like scholars in many other disciplines, have also become much more sensitive to questions of gender and ethnicity. Studies of cultures prior to or beyond modern western Europe have been constrained in the past by lack of sources, or linguistic barriers, but that situation is now beginning to change as web images, new translations, and scholarly commentary are making increasing amounts of source material more readily accessible, intellectually as well as materially. Consequently, the mathematics of the past is no longer regarded simply as a precursor to the mathematics of the present but as an integral part of its own contemporary culture.

As in all thriving academic disciplines these days, those who engage in the history of mathematics are required to cross boundaries. Indeed, one of the great pleasures of working in the subject is that one can learn from the expertise of archaeologists, archivists, sinologists, classicists, orientalists, medievalists, historians of science, linguists, art historians, literary critics, museum curators, and many others. The range of sources has broadened similarly and is no longer restricted to the books or manuscripts that once expounded the latest ideas but includes correspondence, diaries, rough notes, exercise books, measuring instruments, calculating machines, paintings, sketches, diaries, and novels. The last item may sound surprising, but novelists may be the most astute and articulate recorders of contemporary views of mathematics; readers interested in following up this theme will find more under 'Further reading'.

The questions asked by historians over the last 50 years have both changed and diversified. It is no longer enough simply to ask who discovered what and when. We also want to know what mathematical practices engaged groups of people or individuals, and why. What historical or geographical influences were at work? How were mathematical activities perceived, by the participants or by others? What aspects were particularly valued? What steps were taken to preserve or hand on mathematical expertise? Who was paying for it? How did an individual mathematician manage his (or her) time and skill? What were their motivations? What did they produce? What did they do with it? And with whom did they discuss, collaborate, or argue along the way?

Most of the answers to most of these questions will be difficult to find with any degree of certainty. Historians of mathematics, like all other historians, work with sparse evidence, from which they must reconstruct as carefully as possible incomplete stories about the past. The attempt remains worthwhile for what we can learn about a human activity as ancient and as widespread as producing literature or music, and which has manifested itself in as rich a variety of cultural forms: doing and creating mathematics.

Further reading

Chapter 1

Primary sources

Robert Recorde, *The Pathway to Knowledg* (London, 1551); painstakingly reprinted by Gordon and Elizabeth Roberts (TGR Renascent Books, 2009).

Secondary sources

Markus Asper, 'The two cultures of mathematics in ancient Greece', in Eleanor Robson and Jacqueline Stedall (eds), *The Oxford Handbook of the History of Mathematics* (Oxford University Press, 2009), pp. 107–132.

Simon Singh, *Fermat's Last Theorem* (Fourth Estate, 1997; Harper Perennial, 2007).

Benjamin Wardhaugh, 'Mathematics in English printed books, 1473–1800: a bibliometric analysis', *Notes and Records of the Royal Society*, 63(2009): 325–38.

Chapter 2

Primary sources

The Suàn shù shū, writings on reckoning: a translation of a Chinese mathematical collection of the second century BC, *with explanatory commentary*, tr. Christopher Cullen (Needham Research Institute, 2004).

Fibonacci's Liber abaci: Leonardo Pisano's book of calculation, tr. L. E. Segal (Springer, 2002).

Secondary sources

Christopher Cullen, 'People and numbers in early imperial China', in Eleanor Robson and Jacqueline Stedall (eds), *The Oxford Handbook of the History of Mathematics* (Oxford University Press, 2009), pp. 591–618.

G. E. R. Lloyd, 'What was mathematics in the ancient world?', in Eleanor Robson and Jacqueline Stedall (eds), *The Oxford Handbook of the History of Mathematics* (Oxford University Press 2009), pp. 7–25.

Benjamin Wardhaugh, 'Poor Robin and Merry Andrew: mathematical humour in Restoration England', *BSHM Bulletin*, 22(2007): 151–9.

Chapter 3

Primary sources

The thirteen books of Euclid's *Elements* in MS D'Orville 301, from 888, http://www.claymath.org/library/historical/euclid/ last accessed November 2011.

David Joyce, 'A Quick Trip through the Elements', compiled in 2002, http://aleph0.clarku.edu/~djoyce/java/elements/trip.html, last accessed January 2012.

Secondary sources

June Barrow-Green, '"Much necessary for all sortes of men": 450 years of Euclid's *Elements* in English', *BSHM Bulletin*, 21(2006): 1–25.

Annette Imhausen, 'Traditions and myths in the historiography of Egyptian mathematics', in Eleanor Robson and Jacqueline Stedall (eds), *The Oxford Handbook of the History of Mathematics* (Oxford University Press, 2009), pp. 781–800.

Victor Katz (ed.), *The mathematics of Egypt, Mesopotamia, China, India, and Islam: a sourcebook* (Princeton University Press, 2007).

Reviel Netz and William Noel, *The Archimedes Codex: revealing the secret of the world's greatest palimpsest* (Weidenfeld and Nicolson, 2007).

Eleanor Robson, *Mathematics in ancient Iraq: a social history* (Princeton University Press, 2006).

Corinna Rossi, 'Mixing, building, and feeding: mathematics and technology in ancient Egypt', in Eleanor Robson and Jacqueline

Stedall (eds), *The Oxford Handbook of the History of Mathematics* (Oxford University Press, 2009), pp. 407–28.
Benjamin Wardhaugh, *How to read historical mathematics* (Princeton University Press, 2010).

Chapter 4

Primary sources

Copy books in the John Hersee collection owned by the Mathematical Association, in the David Wilson Library at the University of Leicester.

Secondary sources

Marit Hartveit, 'How Flora got her cap', *BSHM Bulletin*, 24(2009): 147–58.
Eleanor Robson, *Mathematics in ancient Iraq* (Princeton University Press, 2008).
Polly Thanailaki, 'Breaking social barriers: Florentia Fountoukli (1869–1915)', *BSHM Bulletin*, 25(2010): 32–8.

Chapter 5

Sonja Brentjes, 'Patronage of the mathematical sciences in Islamic societies', in Eleanor Robson and Jacqueline Stedall (eds), *The Oxford Handbook of the History of Mathematics* (Oxford University Press, 2009), pp. 301–27.

Chapter 6

Primary sources

Euclid, *The first six books of the Elements of Euclid*, beautifully reproduced in colour from Oliver Byrne's 1847 original by Werner Oechslin and Petra Lamers-Schutze (Taschen, 2010).
Euclid, *Elements*, Oliver Byrne's coloured edition of 1847 combined with David Joyce's interactive version of 2002 http://www.math.ubc.ca/~cass/Euclid/byrne.html, last accessed January 2012.

Secondary sources

Glen van Brummelen, 'Filling in the short blanks: musings on bringing the historiography of mathematics to the classroom', *BSHM Bulletin*, 25(2010): 2–9.

Chapter 7

Tony Mann, 'From Sylvia Plath's *The Bell Jar* to the Bad Sex Award: a partial account of the uses of mathematics in fiction', *BSHM Bulletin*, 25(2010): 58–66.

“牛津通识读本”已出书目

古典哲学的趣味
人生的意义
文学理论入门
大众经济学
历史之源
设计，无处不在
生活中的心理学
政治的历史与边界
哲学的思与惑
资本主义
美国总统制
海德格尔
我们时代的伦理学
卡夫卡是谁
考古学的过去与未来
天文学简史
社会学的意识
康德
尼采
亚里士多德的世界
西方艺术新论
全球化面面观
简明逻辑学
法哲学：价值与事实
政治哲学与幸福根基
选择理论
后殖民主义与世界格局
福柯
缤纷的语言学
达达和超现实主义
佛学概论
维特根斯坦与哲学
科学哲学
印度哲学祛魅
克尔凯郭尔
科学革命
广告
数学
叔本华
笛卡尔
基督教神学
犹太人与犹太教
现代日本
罗兰·巴特
马基雅维里
全球经济史
进化
性存在
量子理论
牛顿新传
国际移民
哈贝马斯
医学伦理
黑格尔
地球
记忆
法律
中国文学
托克维尔
休谟
分子
法国大革命
民族主义
科幻作品
罗素
美国政党与选举
美国最高法院
纪录片
大萧条与罗斯福新政
领导力
无神论
罗马共和国
美国国会
民主
英格兰文学
现代主义
网络
自闭症
德里达
浪漫主义
批判理论

德国文学
戏剧
腐败
医事法
癌症
植物
法语文学
微观经济学
湖泊
拜占庭
司法心理学
发展
农业
特洛伊战争
巴比伦尼亚
河流
战争与技术
品牌学
数学简史
儿童心理学
时装
现代拉丁美洲文学
卢梭
隐私
电影音乐
抑郁症
传染病
希腊化时代
知识
环境伦理学
美国革命
元素周期表
人口学
社会心理学
动物
项目管理
美学
电影
俄罗斯文学
古典文学
大数据
洛克
幸福
免疫系统
银行学
景观设计学
神圣罗马帝国
大流行病
亚历山大大帝
气候
第二次世界大战
中世纪
工业革命
传记
公共管理